七巧新譜

嘉慶丙子春月重鐫

聽雨樓珍玩

The title page of a Chinese tangram book (see page 59).

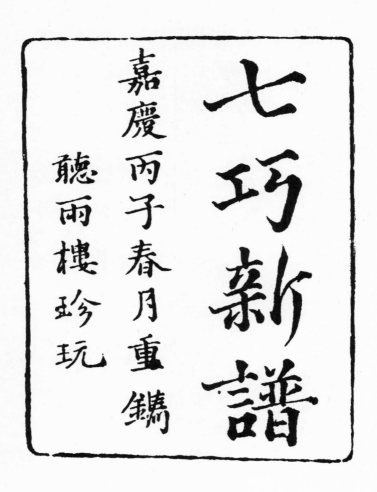

TANGRAMS—

330 Puzzles

by Ronald C. Read

Dover Publications, Inc., New York

Published in Canada by General Publishing Company, Ltd., 30 Lesmill Road, Don Mills, Toronto, Ontario.

Published in the United Kingdom by Constable and Company, Ltd., 10 Orange Street, London W.C.2.

Tangrams—330 Puzzles is a new work, first published by Dover Publications, Inc., in 1965.

Library of Congress Catalog Card Number 65-26651

Manufactured in the United States of America

Dover Publications, Inc.
180 Varick Street
New York, N.Y. 10014

Table of Contents

Introduction

THE seven-piece puzzle, or tangram as it is usually known in the West, has come to us from China. It is in all probability the original "Chinese puzzle"—the proverbial prototype of all that is perplexing and tricky. This puzzle resembles the familiar jigsaw puzzle in that it is concerned with the fitting together of geometrical shapes, yet in all other respects the two puzzles are completely different; for whereas the jigsaw puzzle consists of a large number of pieces of rather complicated shapes, which have to be fitted together in a unique way, the tangram consists of only seven pieces of simple shapes, and the whole charm of the puzzle lies in the extraordinary variety of ways in which these pieces can be put together.

The seven pieces which make up the tangram can be cut from a single square, as shown in Figure 1. There are thus two small triangles, one medium-sized triangle, and two large triangles, in addition to a square and a lozenge-shaped piece.

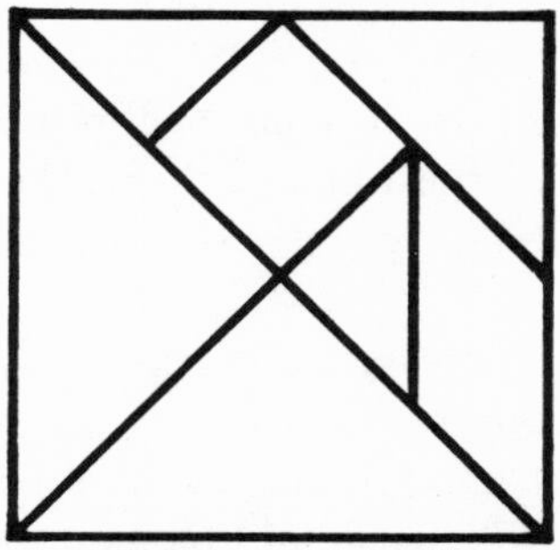

FIGURE 1

The medium-sized triangle and the square and the rhomboid are all twice the area of one of the small triangles; each of the large triangles is four times the area of a small triangle. All the angles in these pieces are either right angles or angles of 45° or 135°. A set of tangram pieces can easily be made from cardboard. However, the enthusiast will probably want to make his set from wood, plastic or some other more durable material.

With a set of seven tangram pieces, the reader can make (if he tries hard enough) all the outlines that are to be found in the following pages except those of the last two chapters, and a quick glance through these pages will reveal the surprisingly large number of different outlines that can be constructed from such a small number of pieces. This is what has made the tangram popular in the country of its origin and also in the West. From time to time it has been revived on both sides of the Atlantic and has enjoyed a certain vogue, even though it has never achieved the status of such established pastimes as checkers or tic-tac-toe.

The tangram has a reputation for being very old, and this may well be, though one should beware the common tendency to ascribe exaggerated age to all things Chinese. I have in my possession two books of tangrams printed in China in 1813 and 1823. In the preface to one of these the author says that "the origin of the seven-piece puzzle is not known . . ." This would suggest that the puzzle was even then regarded as old, though just how old is a matter for conjecture. Despite this, I have been quite unable to find any Chinese books on tangrams, or even any reference to this kind of puzzle, dating earlier than the beginning of the nineteenth century, at which time a large number of books on the subject seem to have been published.

There was, apparently, a resurgence of interest in the tangram in China around this time, resulting in the publication in book form of what may previously have been handed down only by word of mouth. It should be mentioned here that the Chinese have always regarded the seven-piece puzzle as predominantly a game for children and women, and therefore perhaps hardly a worthy subject for books. I venture to hope that this is not true of the present book, which contains many subtleties devised by occidental puzzlers that are not to be found in the Chinese texts.

It was during the first half of the nineteenth century that the tangram became known in Western countries, and books about it started to appear. At first, these books were nothing more than exact copies of Chinese texts, with the addition of hard covers and some sort of English preface explaining—often not very clearly—what the puzzle was about. A copy of one such book, now in the British Museum, bears on its cover the following:

A Grand Eastern Puzzle

THE following Chineze Puzzle is recommended to the Nobility, Gentry and others, being superior to any hitherto invented for the Amusement of the Juvenile World, to whom it will afford unceasing recreation and information: being formed on Geometrical principles, it may not be considered as trifling to those of mature years, exciting interest, because difficult and instructive, imperceptibly leading the mind on to invention and perseverance. —— The Puzzle consists of five triangles, a square, and a rhomboid, which may be placed in upwards of THREE HUNDRED and THIRTY Characters, greatly resembling MEN, BEASTS, BIRDS, BOATS, BOTTLES, GLASSES, URNS, etc. The whole being the unwearied exertion of many years study and application of one of the Literati of China, and is now offered to the Public for their patronage and support.

ENTERED AT STATIONER'S HALL

Published and sold by
C. DAVENPORTE and CO.
No. 20, Grafton Street, East Euston Square.

Later on, publications appeared in Europe and America which were more original, but which still depended very heavily on the earlier sources. The same outlines appear, often in the same order, and often with mistakes (common in the Chinese books) faithfully copied! One such publication which I have in my possession was printed in Philadelphia in 1844. Its derivation from the earlier Chinese books is quite apparent, but it is greatly inferior to them. Approximately three hundred tangrams are given, but they are jumbled together any which way, and no indication is given of what the outlines are supposed to represent. This reduces the puzzle to the mere construction of meaningless geometrical patterns, which must have provided rather dull entertainment.

Lewis Carroll had in his library a book of tangrams entitled *The Fashionable Chinese Puzzle*, which seems to have been very similar to the American publication just mentioned. The great English puzzler H. E. Dudeney, who later acquired this book, describes it as follows:

> It contains three hundred and twenty-three Tangram designs, mostly nondescript geometrical figures, to be constructed from the seven pieces. . . . There is no date, but the following note fixes the time of publication pretty closely: "This ingenious contrivance has for some time past been the favourite amusement of the ex-Emperor Napoleon, who, being now in a debilitated state and living very retired, passes many hours a day in thus exercising his patience and ingenuity."

This reference to Napoleon's interest in the puzzle occurs also in the preface to the Philadelphia publication, but it would seem unlikely that there is any truth in the assertion. Napoleon's biographers, writing of his exile on St. Helena, mention that he was interested (though not skilled) in billiards, chess and *reversi*; nothing is said of anything at all resembling the tangram. Still, it is not impossible that Napoleon was acquainted with this form of entertainment, and the story of the Emperor whiling away his declining years in this manner, if not true, is at all events a pleasing enough fabrication.

All this goes to show that historical information about the tangram is almost nonexistent; but though fact is in short supply there is fiction in abundance. A detailed and highly colored "tangram legend" has come into existence, mainly due to the efforts of Sam Loyd, that great American puzzle expert of the turn of the century, aided and abetted by H. E. Dudeney in England. Sam Loyd's contribution to the literature of tangrams (and a very notable one it is, too) was a book called *The 8th Book of Tan*, which was published by Loyd himself in 1903. In this book Sam Loyd brought his own peculiar genius to bear on all aspects of the puzzle; in addition to creating hundreds of completely new shapes, he also dreamt up some paradoxical twists and quirks that had never been thought of before.

Along with these tangrams Sam Loyd supplied a running commentary, in which he gave what has every appearance of being a remarkably precise and carefully documented account of the history of the puzzle, its esoteric religious significance and symbolism, and its relation to the theorem of Pythagoras and other parts of Euclid's geometry, not to mention much additional scholarly and erudite information.

In point of fact, however, this painstaking work of scholarship is one delightful spoof from beginning to end! One example will suffice to show the sort of tall yarn that Sam Loyd pulls out of the hat. He quotes the researches of the "famous Professor Challenor" (a purely fictitious character, as far as I have been able to discover), who is said to have made a special study of the seven-piece puzzle. According to these researches the puzzle originated with a monumental work in seven volumes entitled *The Seven Books of Tan*, which was compiled in China some four thousand years ago. These books are described as "rare" (an understatement), and mention is made of one of them, printed in gold leaf upon parchment, that was found in Peking. Now, none of this is even remotely plausible. In 2000 B.C. the Chinese were still at the stage in which their literary achievements mainly consisted of crude pictographs scratched on tortoise shells; they had certainly not progressed even to writing on thin strips of bamboo—the way in which the Confucian classics, for example, were originally recorded (around 500 B.C.). A work of seven volumes would be quite out of the question at that time, and a book printed in gold leaf on parchment would be most unusual (to say the least) at any time.

I must beg the reader's indulgence if he considers that this is all too obvious to be worth the trouble of explanation. I take the trouble because it would seem that H. E. Dudeney accepted these tall stories at their face value; if he did not, he at least repeated them with such seriousness and seeming credence that his readers, faced with the unanimity of two famous authorities, might well believe them to be true.

This, of course, is exactly what Sam Loyd would like. His spoofing is clearly deliberate, and is completely in keeping with the good-natured leg-pulling that lends special appeal to so many of his puzzles.*

Dudeney's long preamble to Problem 169 (A Tangram Paradox) in his *Amusements in Mathematics*† repeats the summary of Professor Challenor's researches and describes the attempts (by the eminent philologist Sir James Murray) to check on this story. Needless to say, the result was that the legendary "Tan" and his seven books are completely unknown in China. Dudeney then goes on, apparently in all innocence, to contribute an additional piece of spurious mystification to the "tangram legend." He refers to a Chinese book of tangrams about which an American correspondent has written to him, and he reproduces a Chinese inscription from its first page. His reason for doing this is that—

> The owner of the book informs me that he has submitted it to a number of Chinamen in the United States and offered as much as a dollar for a translation. But they all steadfastly refused to read the words, offering the lame excuse that the inscription is Japanese. Natives of Japan, however, insist that it is Chinese. Is there something occult and esoteric about Tangrams, that it is so difficult to lift the veil?

* Two volumes of Loyd's puzzles have been republished recently, in which this propensity for fooling the reader is amply demonstrated. I refer to *Mathematical Puzzles of Sam Loyd* (two volumes), edited by Martin Gardner. New York: Dover Publications, Inc.

† Reprinted by Dover Publications Inc.

FIGURE 2

Now it so happens that I have a copy of this same book. The "mysterious inscription" is shown in Figure 2, along with the rest of the page on which it occurs. First of all we should remark that Dudeney, or his correspondent, has mistaken the back of the book for the front—a common error, since Chinese books start with what we would consider to be the last page. The inscription is actually on the last page but one. It is very badly printed, but otherwise there is nothing particularly difficult about it. One can only suppose that the people who were asked to translate it were less well versed in Chinese (or Japanese) than the owner of the book thought. The inscription reads roughly as follows: "Two men facing each other drinking. This shows the versatility of the seven-piece puzzle."

Clearly, this is simply a caption for the tangram outlines which appear beneath it. Far from being mysterious or occult, the inscription is not even out of the ordinary, since *every* outline in the book has a caption! What a shame! Another lovely theory ruined by hard facts!

Despite the popular proverb, truth is seldom stranger than fiction, and the full history of the tangram puzzle is probably not nearly as colorful as the legend that Sam Loyd and Dudeney have woven around it. But this in no way detracts from the enjoyment that we can derive from this pastime, as the following pages are intended to show. The object of the game is, of course, to put the seven tangram pieces together so as to form the various outlines given throughout the book. Some of these are quite easy, others less so. If an outline really stumps you, look up the solution at the back of the book. But the Solutions should be your last resort; don't consult them unless you are really stuck. You may want to try to construct some original tangrams, and some interesting possibilities are suggested here and there in the text.

Finally, a word about the name "tangram" itself. The "gram" part presents no difficulty; it is a common ending denoting something written or drawn (as in "diagram," for example). The origin of the first part of the word is more in doubt. There are several explanations, but one is so much more plausible than any other as to be almost certainly correct. The Tang dynasty was one of the greatest in Chinese history; so much so that in certain South Chinese dialects the word *tang* (or *tong*) is synonymous with "Chinese." Now, a European visiting China in the first half of the nineteenth century, if he learned any Chinese at all, would be more likely to learn a Southern dialect than any other. Thus a traveler bringing the puzzle from China to the West, and wanting a name for it, would be very likely to take the word *tang*—that is, Chinese—and combine it with the familiar ending "gram." Dudeney quotes Sir James Murray as being of the opinion that the name was probably introduced in this way by an American some time between 1847 and 1864, the latter date marking the word's first appearance in Webster's Dictionary.

If this conjecture about the derivation of the name tangram is correct, then the tangram is truly the "Chinese puzzle," in name as well as in origin.

1. *Letters and Numbers*

THE seven tangram pieces can be fitted together in such a vast number of ways that in preparing a selection of them it is very difficult to know where to start. Probably as good a way as any is to go right back to our early school days, back to the days when we first learned the alphabet. This time we shall make the letters out of the tangram pieces instead of with pencil and paper.

In the following pages you will find tangram outlines for all the letters of the alphabet. See if you can make them all yourself before looking at the solutions which begin on page 94.

Reconstructing tangram outlines that have been obtained by others is only half the fun that can be derived from this puzzle; as much entertainment, if not more, can be obtained by inventing new outlines. The reader can try his hand at this right away by trying to improve some of the outlines given for the letters. Certainly several could stand improvement; the N, Q and X, in particular, are nothing like as satisfactory as one would wish. The outlines given are the best that I have been able to concoct, but the reader may well be more successful.

For good measure, the digits 1, 2, up to 8 have also been given. Since zero is the same as the letter o, and 9 is simply a 6 upside down, these two digits are not new outlines, and have not been included.

1

2

3

4

5

6

7

8

9

10

11

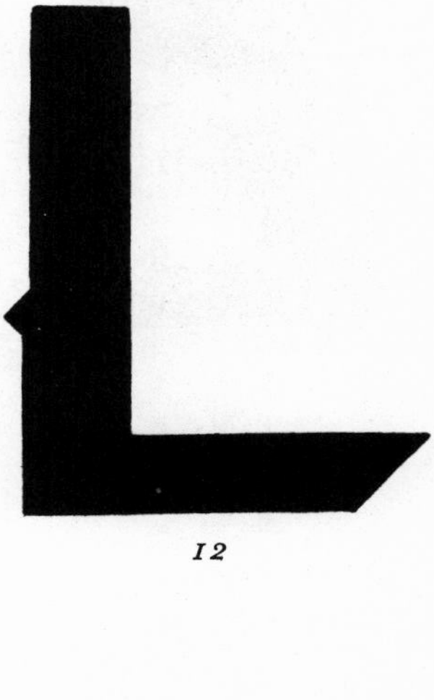

12

13

14

15

16

17

18

19

20

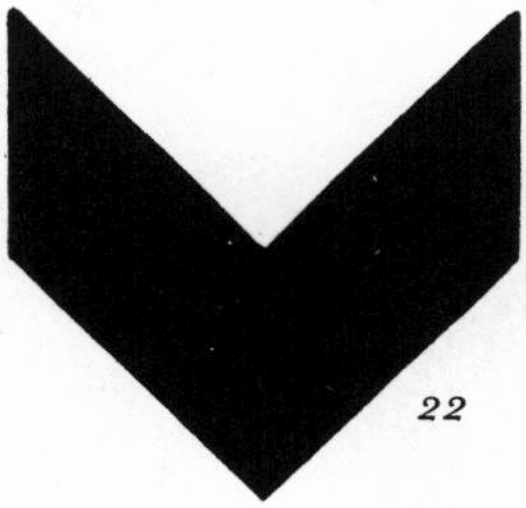

22

21

23

24

26

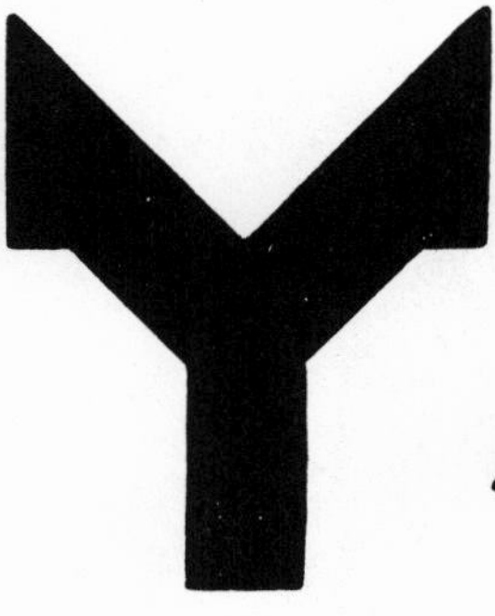

25

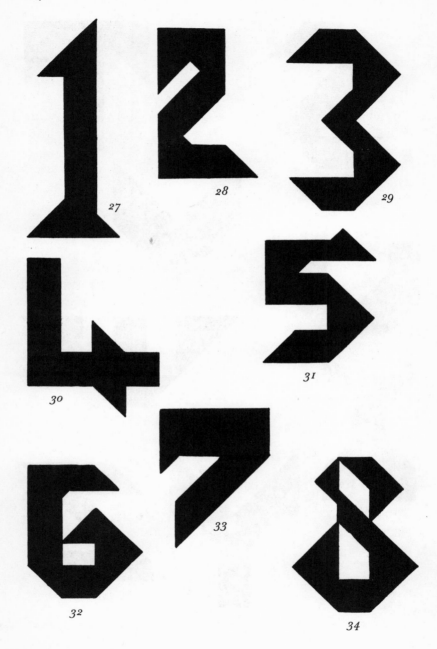

27

28

29

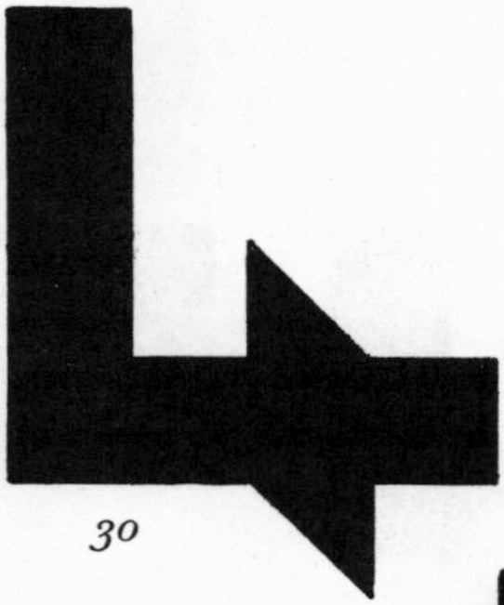

30

31

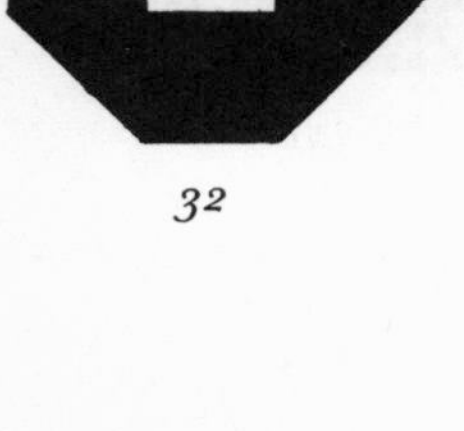

32

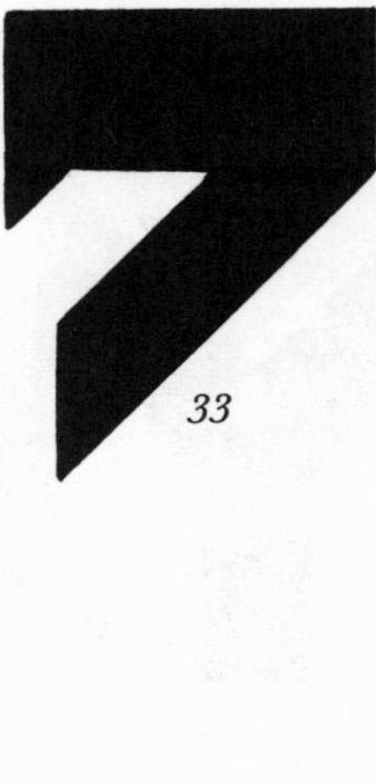

33

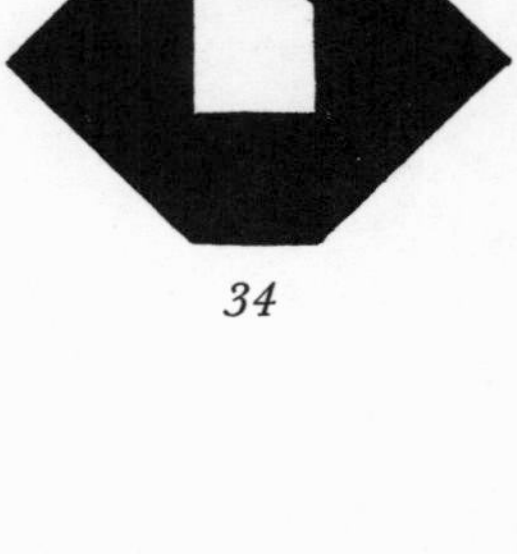

34

2. *Animals*

THE Chinese tangram books, and the later Western books all contain large numbers of outlines representing animals. This book will be no exception, and our contribution to the tangram menagerie lies in the next few pages. Some of the outlines which follow are taken, or adapted, from the Chinese books, and some from Sam Loyd's book. As far as I know, the others are entirely new.

It might be though that the tangram pieces, with their sharp angles and straight sides, could not possibly portray the complicated curved shapes of living creatures, but it is truly surprising how a well-designed tangram can suggest curves where none exist, and complexity where there is only simplicity. See, for instance, how a single piece suffices to suggest the back-curving horns of the mountain goat (No. 35); notice the stratagem by which the thin legs of the stork (No. 50) have been portrayed by three tangram pieces which, individually, are much thicker; note how the careful arrangement of the pieces unerringly conveys the fright of the startled cat (No. 41), the haughtiness of the camel (No. 47) and the graceful lines of the shark (No. 78). We shall see many examples throughout the book of outlines which, by their careful construction, convey much more to the mind's eye than is actually there.

In addition to the animals that have already been mentioned we have:

36. A horned animal of some kind. (Let's not inquire too closely into its precise genus and species!)

37. A polar bear.

38. A giraffe.

39 to 44. Assorted cats, closely followed (naturally) by

45 and 46. Two dogs.

48. A squirrel.

49 to 61. Birds of various shapes and sizes.

62 to 65. Four sinister-looking vultures.

66 to 68. Three horses. Also No. 74.

69 to 72. Four different kinds of bats.

73. A kangaroo.

75. A crocodile.

There follow several sea creatures, mainly fish of various kinds, but also

79. A lobster.

81. A turtle.

82. A seal.

83. A shrimp.

THE TANGRAM ZOO

35

GOATS

36

Polar Bear

37

38 Giraffe

40

39

CATS

41

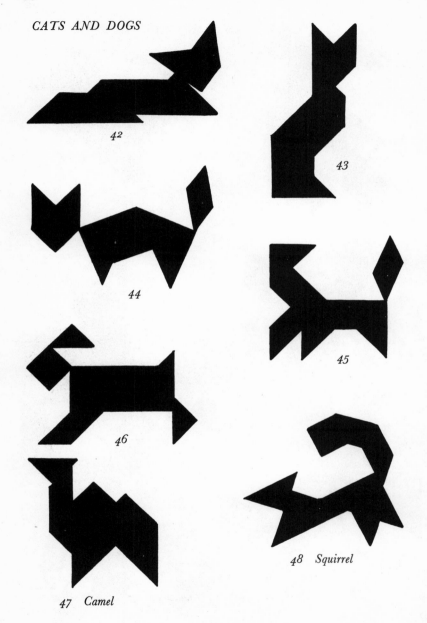

CATS AND DOGS

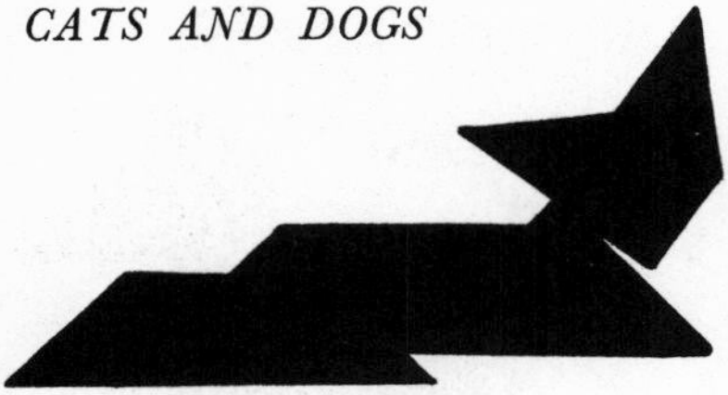

42

43

44

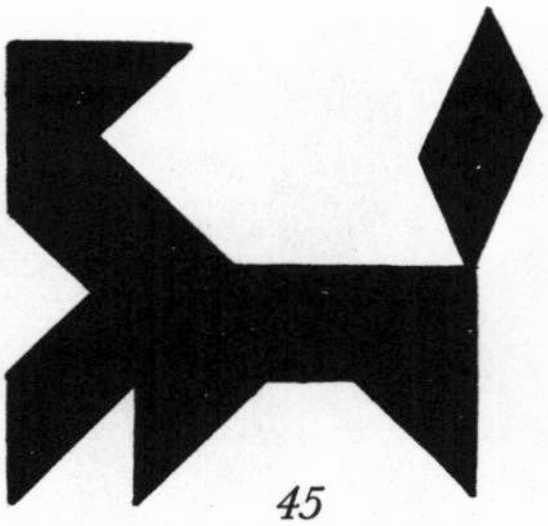

45

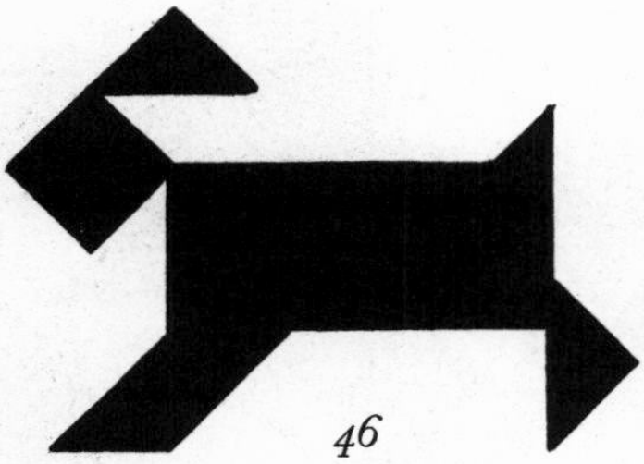

46

47 *Camel*

48 *Squirrel*

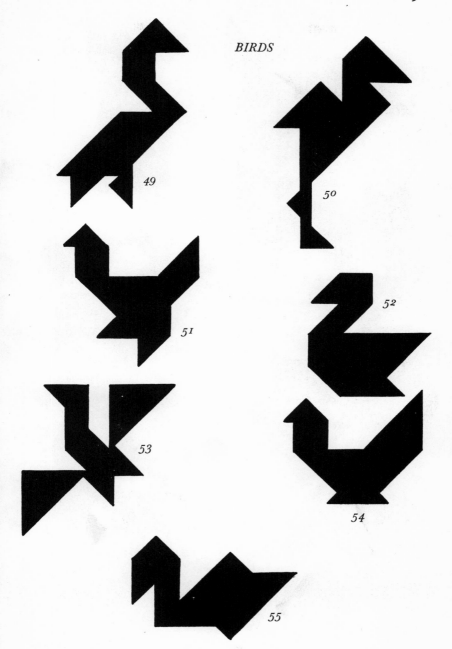

BIRDS

49

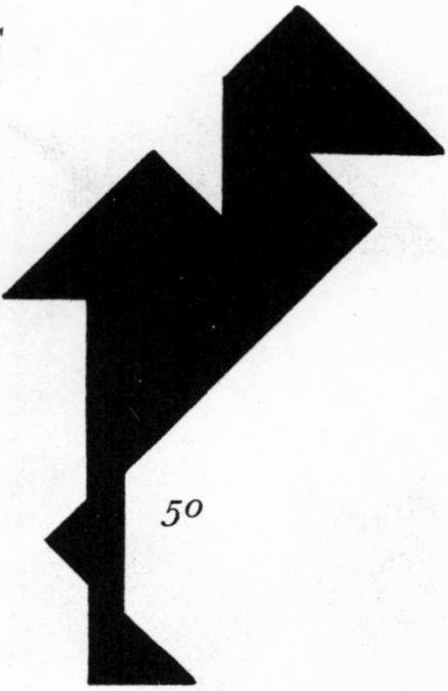

50

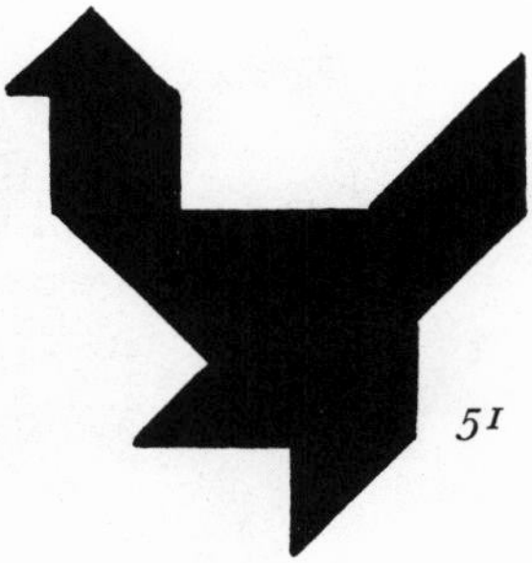

51

52

53

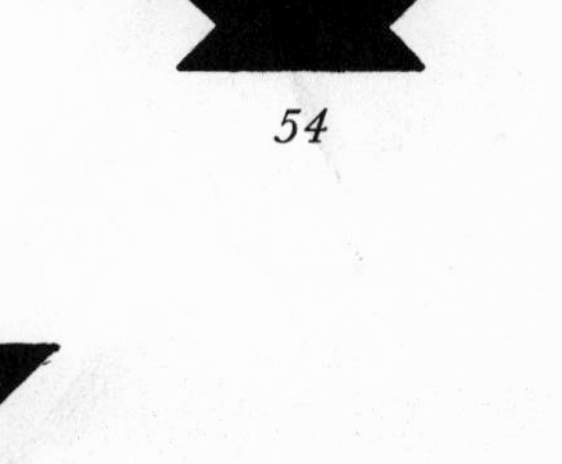

54

55

MORE BIRDS

56

57

58

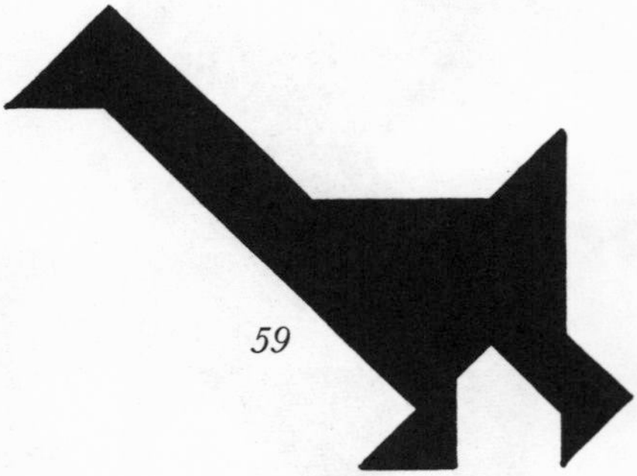

59

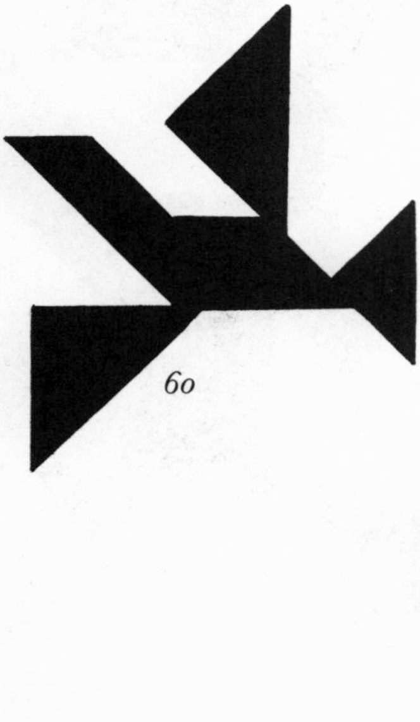

60

61

62

63

VULTURES

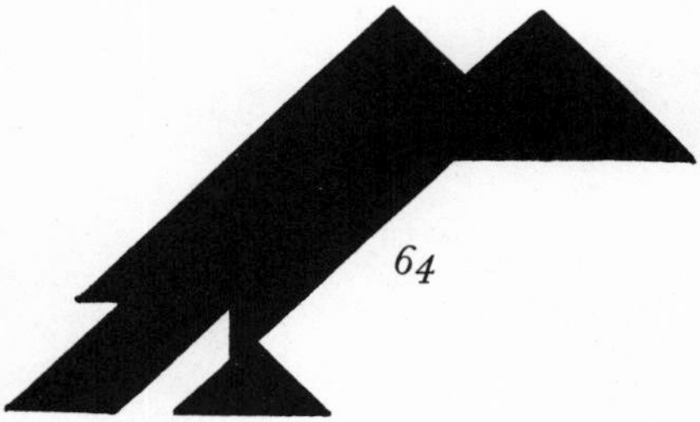

64

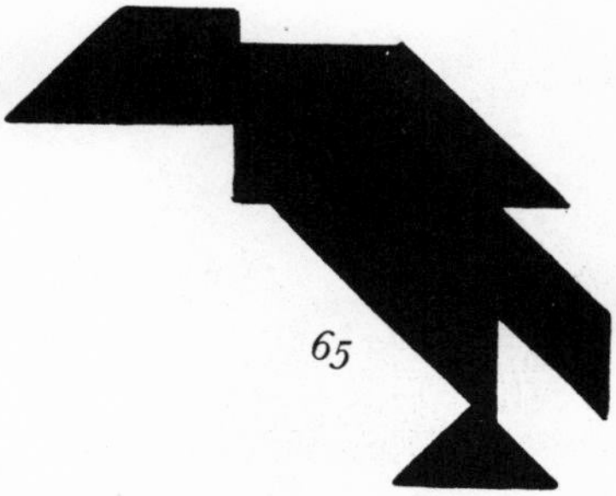

65

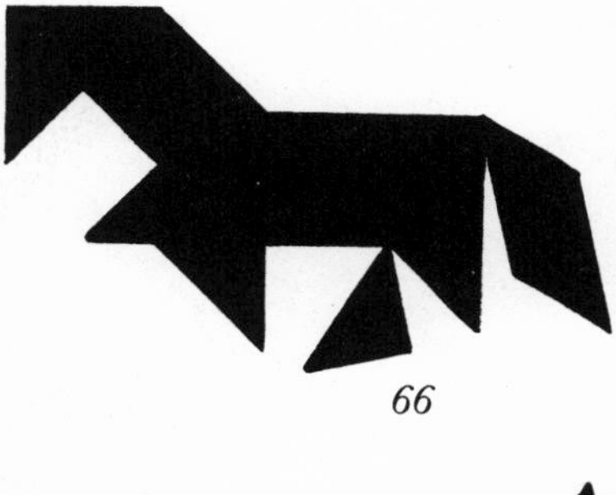

66

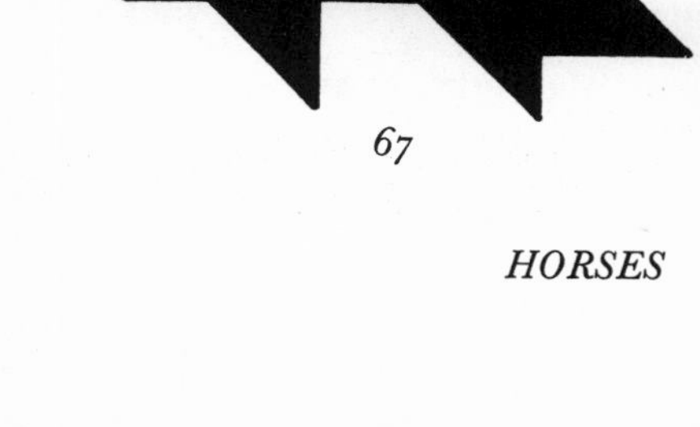

67

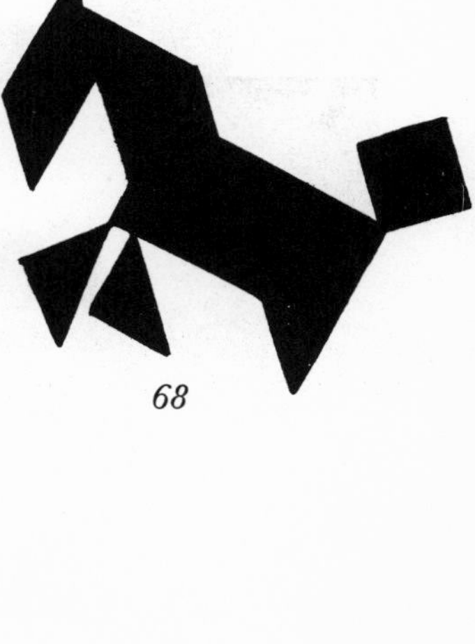

68

HORSES

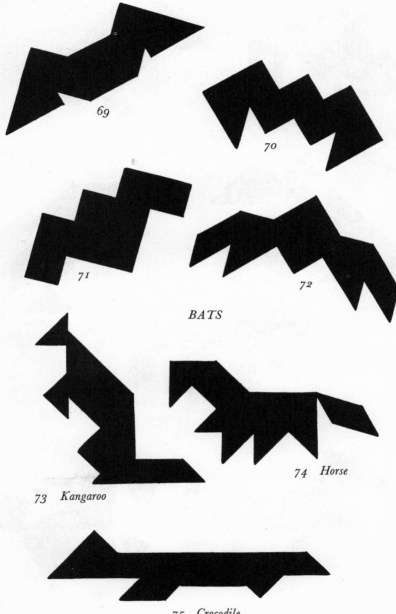

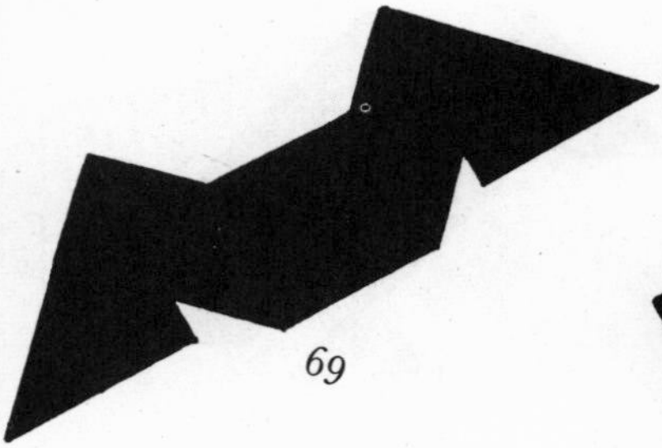

69

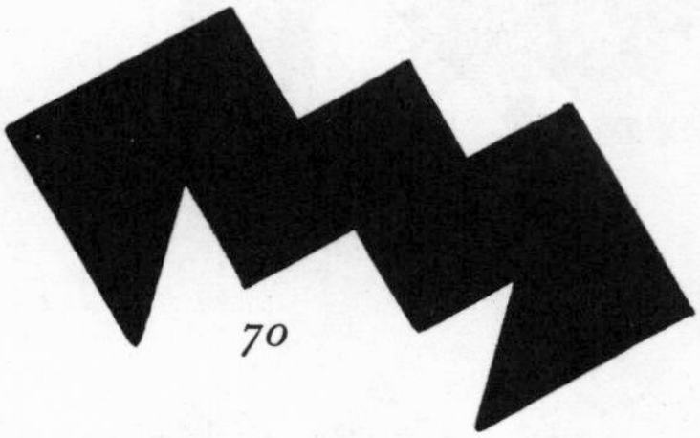

70

71

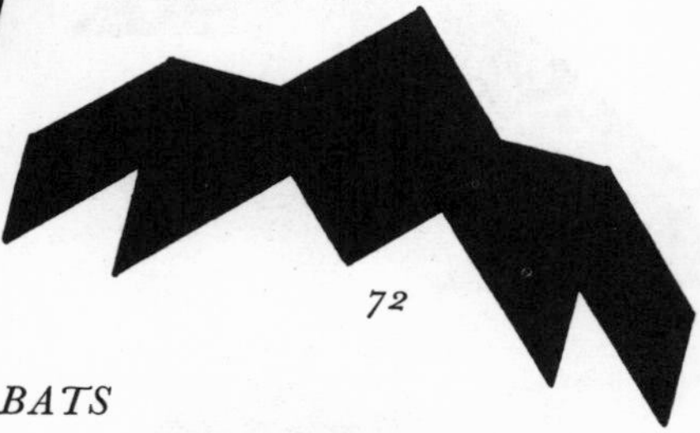

72

BATS

73 *Kangaroo*

74 *Horse*

75 *Crocodile*

SEA CREATURES

76

77

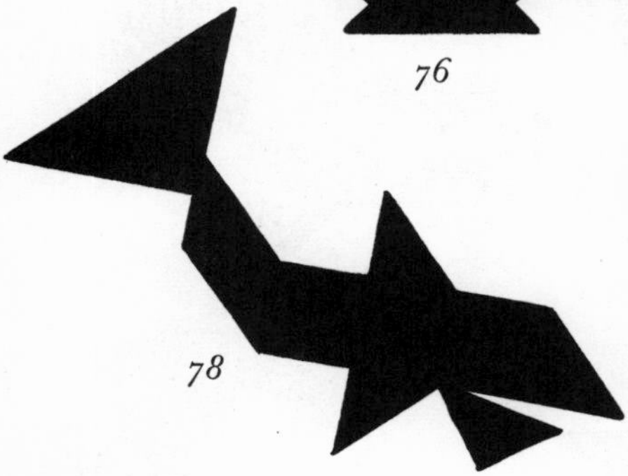

78

79 Lobster

80

81 Turtle

82 Seal

83 Shrimp

SEA CREATURES

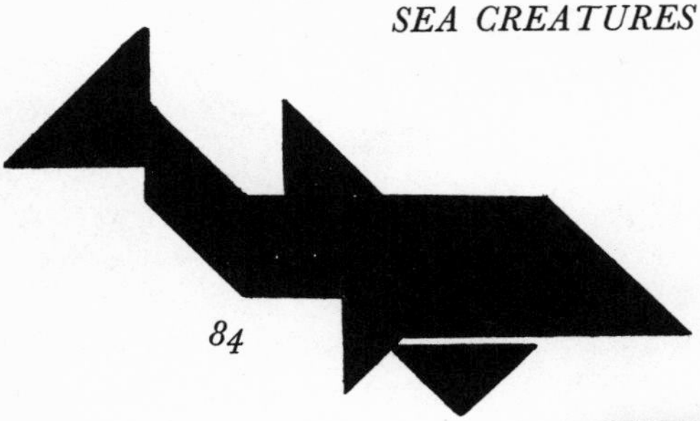

84

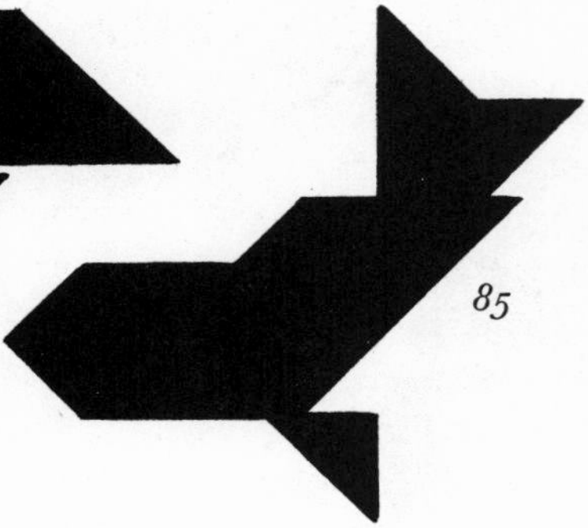

85

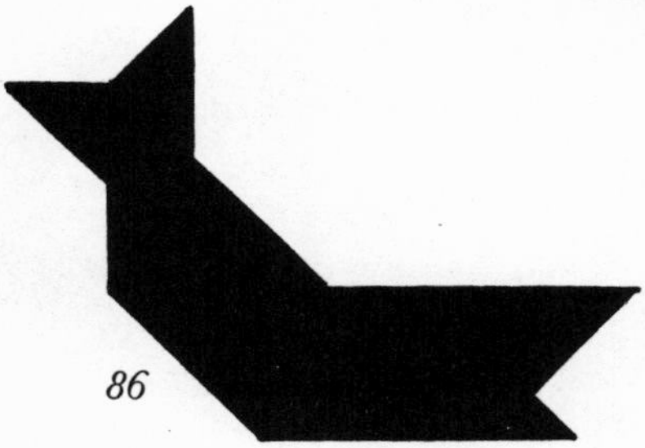

86

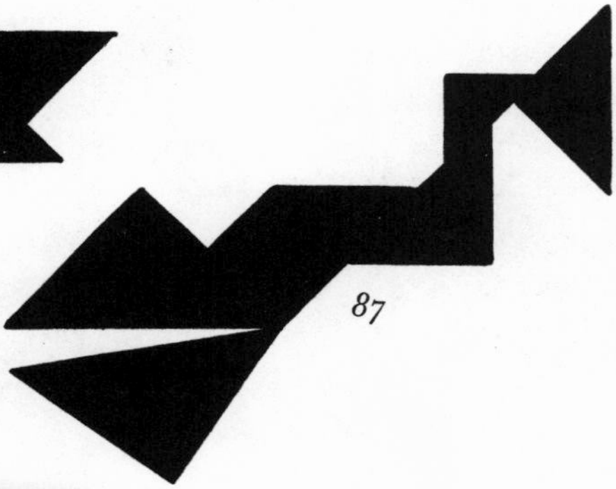

87

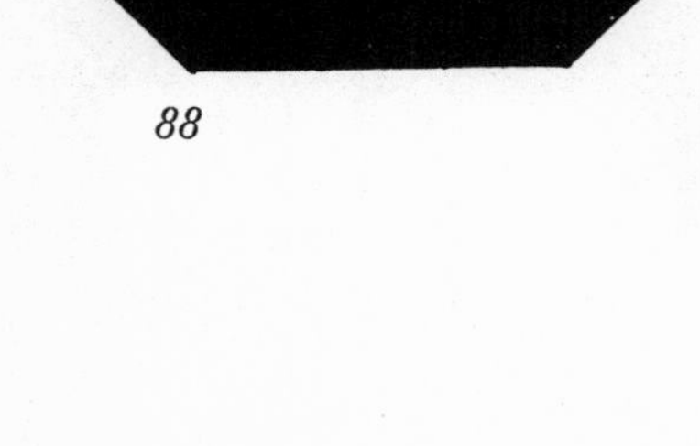

88

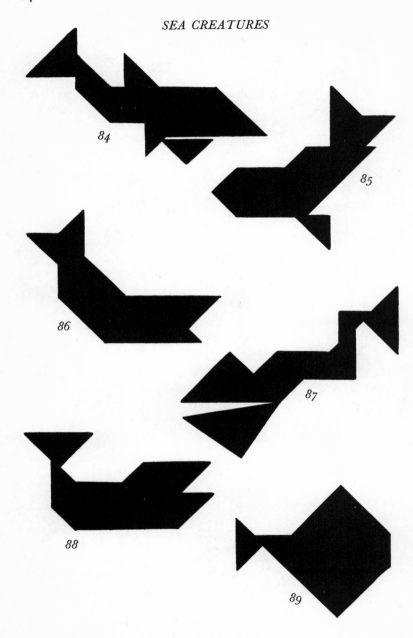

89

3. *Mankind*

While the evolutionist will not deny the propriety of dealing first with animals, we should delay no longer in looking at those tangrams which depict man and his activities. These are among the most interesting of tangrams: for since the human shape is more familiar to us than that of any other object in the universe, we can readily appreciate a tangram which succeeds in portraying the essence of some human form or posture; and just as readily detect the fault in those that do not manage to suggest what they are meant to represent.

Our human parade starts with a shoeshine boy and a customer (Nos. 90 and 91), two men very excited about something, and three acrobats (94, 95, 96), one of whom has just had a fall (or perhaps he is just taking a rest). On the next page are five men who are obviously in a great hurry and a sixth, strolling along with his hands in his pockets, who is clearly in no rush to get anywhere (97 to 102). Then we have four stately medieval ladies (103 to 106) accompanied by a servant (107) and a lady's maid (108).

Next we have "Sam Loyd's Portrait Gallery," twenty-six cleverly constructed heads taken from Sam Loyd's book, plus two more given by Dudeney. Sam Loyd gave names to some of these outlines, as follows:

120. Old Scotch Piper.
121. French Grenadier.
122. Colonial General.
123. A Turk.
124. Aunt Betsy.
125. Uncle Rhube.
126. Mary Smith.
128. John Knox.

129. Tom Sharkey.
130. The Professor.
131. Buffalo Bill.
132. "The Easy Boss."

The Indian chief and his squaw (137 and 138) are also from Sam Loyd's book.

Finally we have four horsemen (139 to 142), a runner (143), and a lady and a gentleman drinking a toast (144 and 145). (Or is he a gentleman? He seems to be seated while the lady is standing—and seated on the floor at that!)

90

91

Shoeshine Boy and Customer

92

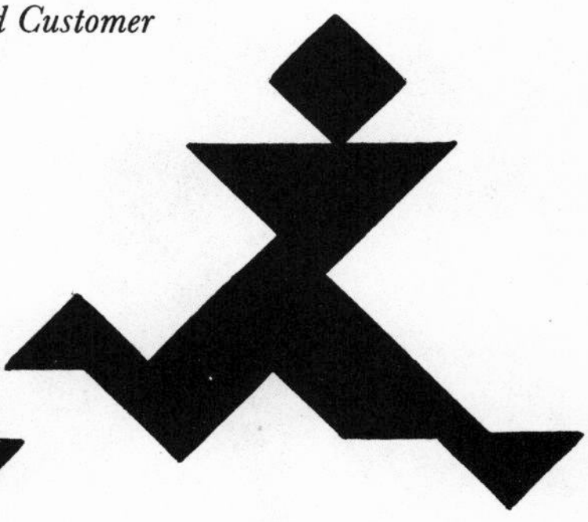

93

94

ACROBATS

95

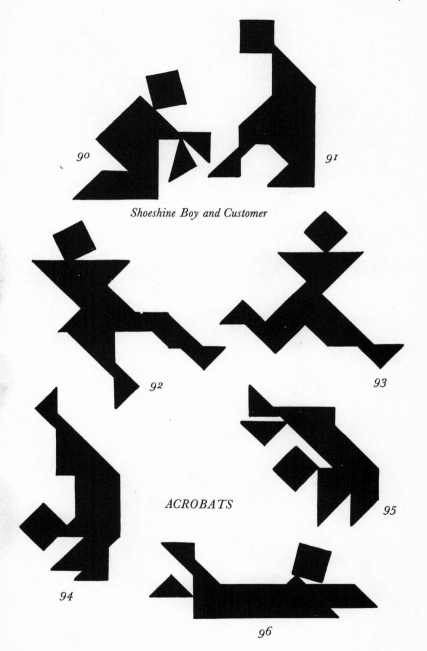

96

97

98

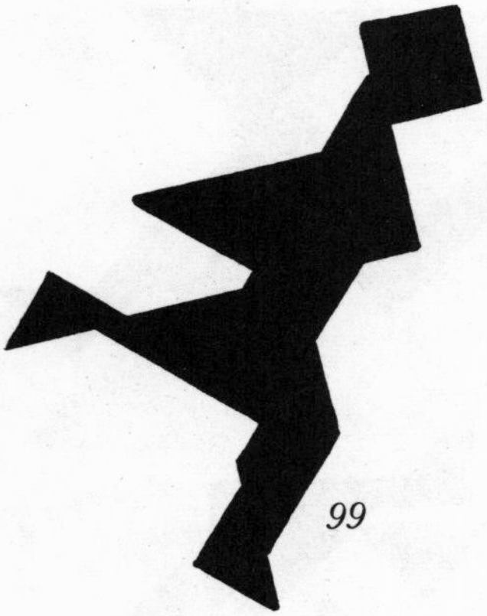

99

100

101

102

MEDIEVAL LADIES

103

104

105

106

AND SERVANTS

107

108

SAM LOYD'S PORTRAIT GALLERY

109

110

111

112

113

114

115

116

117

118

119

120

121

122

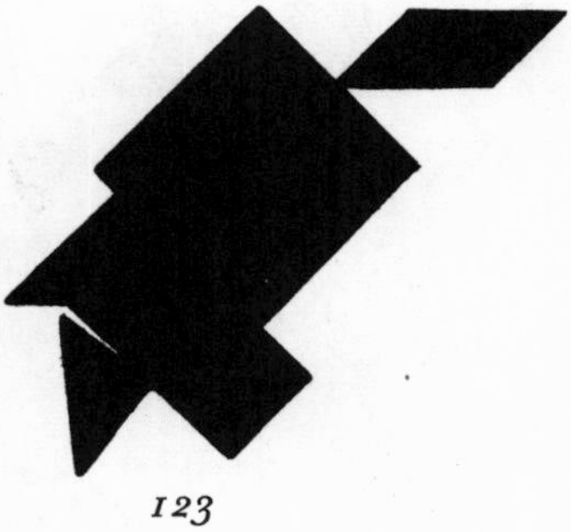

123

124

125

126

127

128

129

130

131

132

133

134

135

136

137

138

HORSEMEN

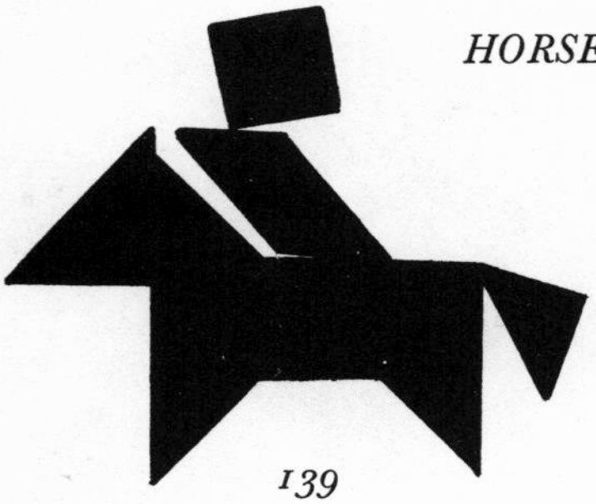

139

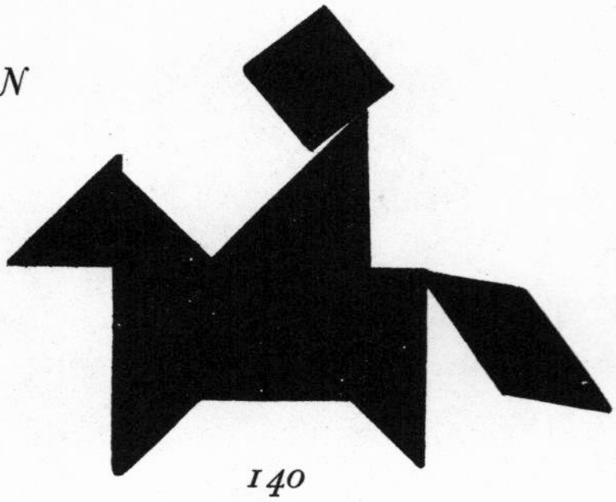

140

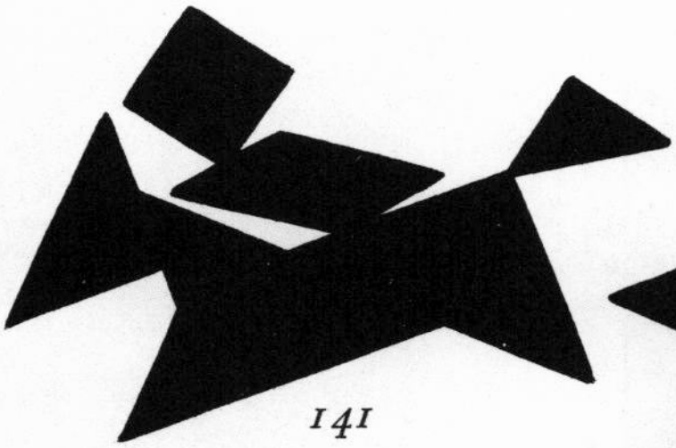

141

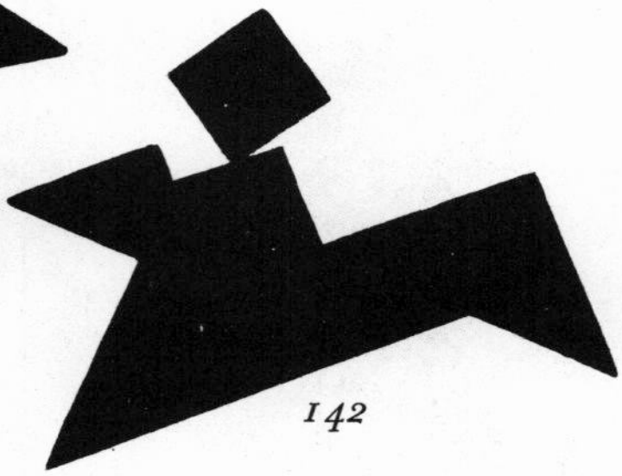

142

143

144

145

4. *Around the House*

IT is only fitting that after looking at man and his activities we should dwell for a few moments on the things that he uses every day. Starting with his pipe (No. 146) we go through a motley collection of shoes, chairs and general odds and ends, ending up with a watering can and a pistol (Nos. 161 and 162). We then have three different kinds of baby carriages—all the latest models and very comfortable, despite the fact that they have square wheels (to which tangram babies have long grown accustomed!). Number 165 is a very superior model which comes complete with a nursemaid (No. 166) to push it. Finally we have Number 167, which is a—now what on earth was that thing meant to be? A saw? A nutmeg-grater? Well, no matter; it is an interesting shape to construct.

146 Pipe

147

148 Candle

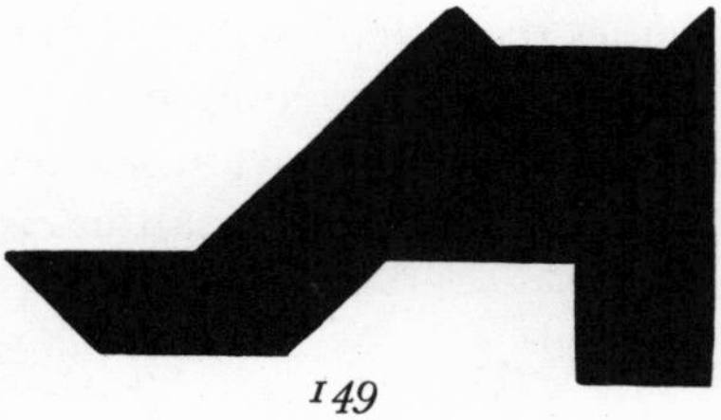

149

150

151

152

SHOES

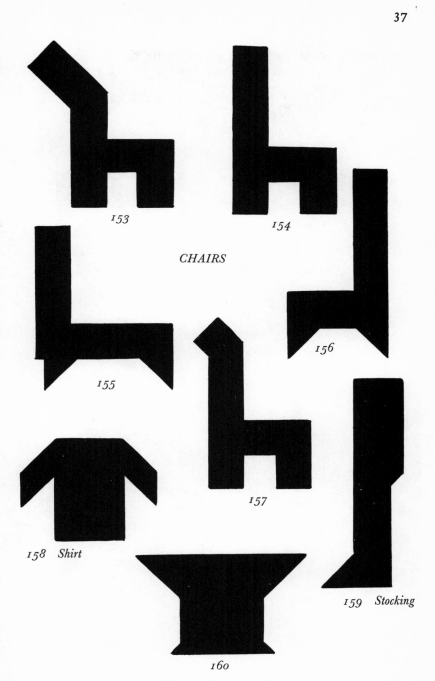

153

154

CHAIRS

155

156

157

158 Shirt

159 Stocking

160

161 Watering can

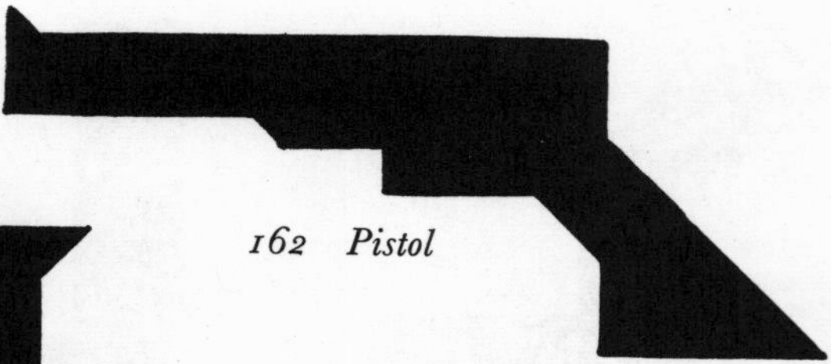

162 Pistol

163

164

BABY CARRIAGES

165

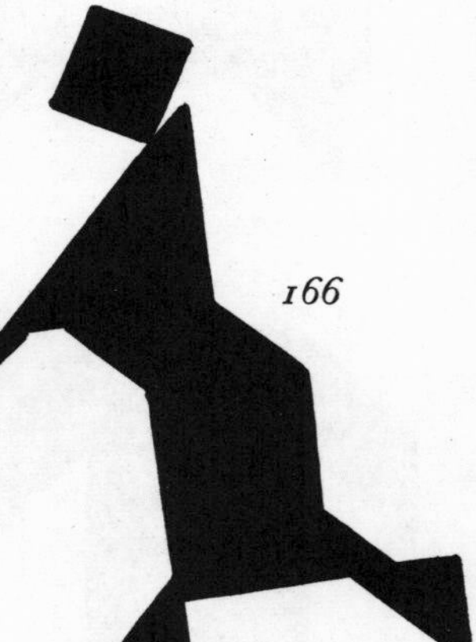

166

167 ???

5. *Boats and Bridges*

THE Chinese books which served to introduce the tangram to the Western world were published in the Treaty Ports of South China, and their pages are full of representations of ships, boats, junks and many other things connected with the sea. It is a curious fact that sailing boats of various kinds seem to be comparatively easy to depict using the seven tangram pieces, notwithstanding the rather complicated silhouettes which they present; whereas steamships, which with their more angular shapes might seem much more suitable for representation by tangrams, are in fact very difficult to concoct. I have been quite unable to produce a convincing tangram of any sort of mechanically propelled craft, with the exception of the launch given in Number 186 (and I won't argue with the reader if he says that this tangram, too, is not very convincing).

Here, then, is something on which the reader can exercise his creative ability; to produce a reasonably good outline of some kind of steamboat or other modern sea-going craft. It will not be easy!

Along with the ships and boats is given an assortment of bridges, mostly taken from the Chinese books. The last two tangrams of this chapter show a lighthouse and the lighthouse keeper being rowed out to it. (Or maybe it is just a sailor taking his maiden aunt for a row around the bay.)

BOATS

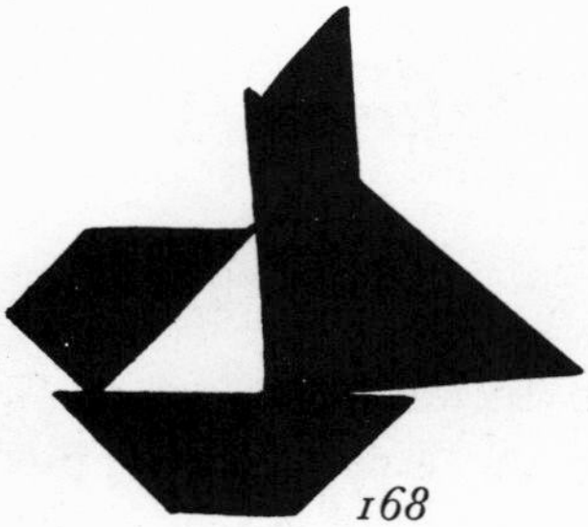

168

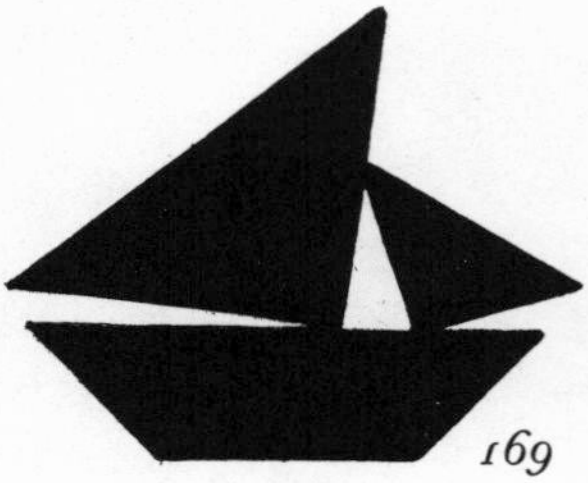

169

170

171

172

173

AND BRIDGES

174

175

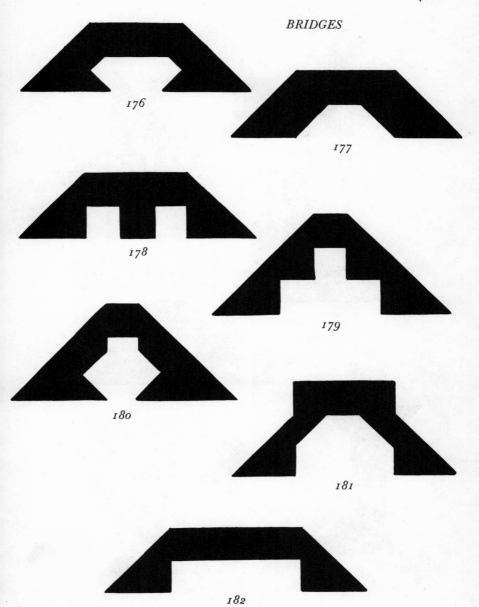

BRIDGES

176

177

178

179

180

181

182

BOATS

183

184

185

186

187 & 188 Rowing to the lighthouse

6. *Stories and Pictures*

Two innovations introduced by Sam Loyd and H. E. Dudeney to extend the scope and interest of the tangram were to construct a series of outlines to illustrate a story, and to bring together several tangram outlines to make a more detailed picture. In this chapter we have two illustrated stories by Sam Loyd, and two pictures by Dudeney.

The first set of tangrams illustrates the well-known nursery rhyme of "The House That Jack Built." In outlines 189 to 199 we see

> . . . the farmer sowing his corn,
> That kept the cock that crowed in the morn,
> That waked the priest all shaven and shorn,
> That married the man all tattered and torn,
> That kissed the maiden all forlorn,
> That milked the cow with the crumpled horn,
> That tossed the dog,
> That worried the cat,
> That killed the rat,
> That ate the malt,
> That lay in the house that Jack built.

To fill out the page, three outlines are given, representing animals that would very likely be found in the vicinity of "The House That Jack Built."

The second story (adapted from Sam Loyd's version) is that of Cinderella. We see Cinderella crying in front of the fireplace (204 and 203); the two ugly sisters (205 and 206), and the fairy godmother with the pumpkin that became a coach and two of the rats that became coachmen (207 to 210). Cinderella dances with the prince (211 and 212), heedless of the clock (213) which is about to strike twelve. The episode of the slipper (214) is too well known to need repetition; everyone knows the story ends with wedding bells (215 and 216) for Cinderella and her prince.

One outline that clearly ought to have been included in this sequence is Cinderella's coach (when it wasn't being a pumpkin). Sam Loyd does not give one in his book, and I have been unable to devise a satisfactory tangram coach; they all turn out looking like baby carriages! Here again the reader may be more successful. By way of compensation I have included Cinderella's coach among the double tangrams later on in this book.

On the next page we have a picture by H. E. Dudeney entitled "A Game of Billiards." Dudeney says, "The players are considering a very delicate stroke at the top of the table."

Dudeney's second picture consists of nine outlines (221 to 229) and will not fit on a single page of this book, so has been spread over two. The description of it is best left to Dudeney himself.

> My second picture is named "The Orchestra", and was designed for the decoration of a large hall of music. Here we have the conductor, the pianist, the fat little cornet-player, the left-handed player of the double-bass, whose attitude is life-like, though he does stand at an unusual distance from his instrument, and the drummer-boy, with his imposing music-stand. The dog at the back of the pianoforte is not howling; he is an appreciative listener.

THE HOUSE THAT JACK BUILT

Illustrated by Sam Loyd

189

190

191

192

193

194

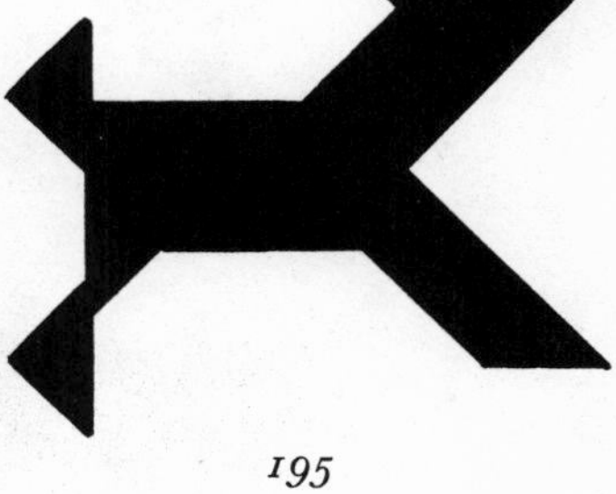

195

196

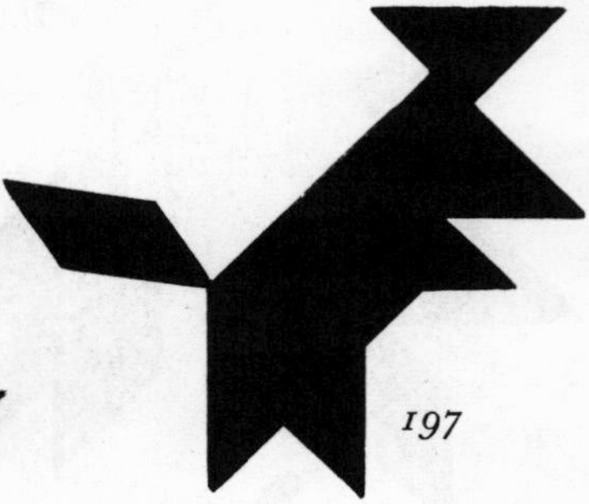

197

198

199

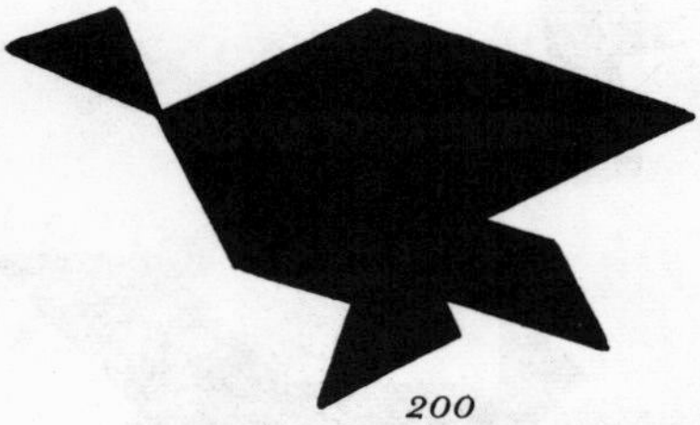

200

201

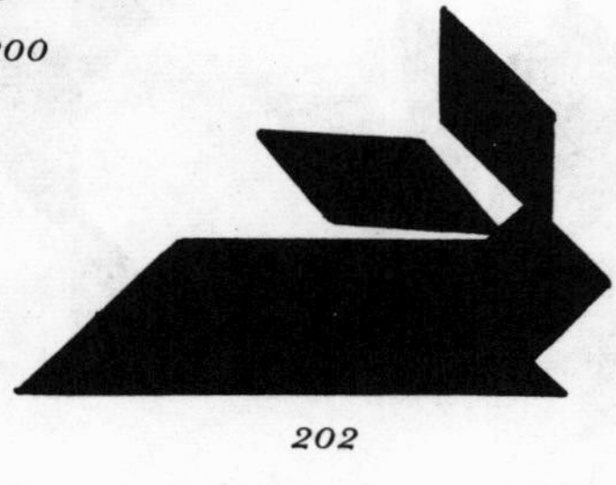

202

THE STORY OF CINDERELLA

Illustrated by Sam Loyd

203

204

205

206

207

208

209

210

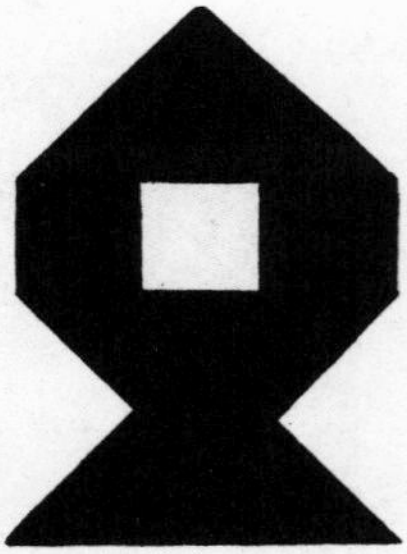

213

211

212

214

215

216

A GAME OF BILLIARDS

By H. E. Dudeney

217

218

219

220

221

222

223

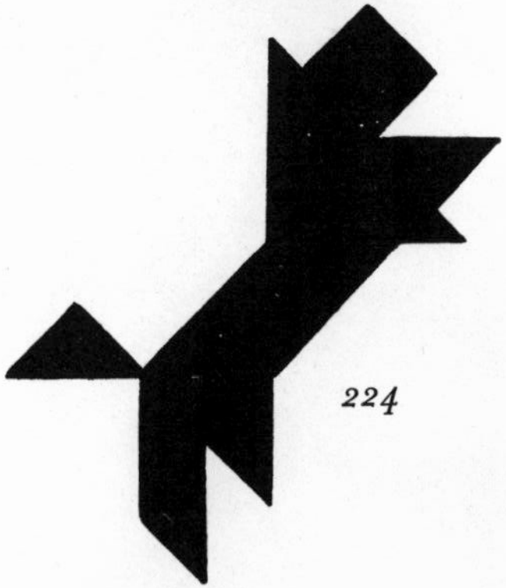

224

225

THE ORCHESTRA

By H. E. Dudeney

226

227

228

229

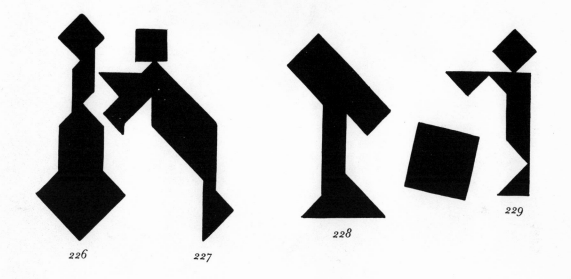

7. *A Little Mathematics*

(Don't be alarmed! You will need very little knowledge of mathematics to follow this chapter.)

As we have already remarked more than once, the number of tangrams is very large. This is really an understatement, for the number is infinite! This we can easily see by looking at Number 229, for example. The bottom corner of the square piece that represents the head of the drummer-boy can touch the rest of the tangram at an unlimited number of points along the line that represents the shoulder and the outstretched arm. It is true that the different outlines that one would get by putting this piece in the different positions would not be *very* different each from the other, but in the strictest sense they would have to be counted as distinct.

It is a little quixotic, however, to accord the same weight to minute variations in a tangram outline as to the differences between two totally distinct outlines, and one naturally wonders whether, by ignoring trivial variations, one could ask the question, "How many tangrams are there?" with some hope of getting a clear-cut answer. Alternatively one might ask the question, "How many tangrams are there of such-and-such a particular kind."

In 1942, two Chinese mathematicians, Fu Tsiang Wang and Chuan-Chih Hsiung, asked, and answered, the question, "How many convex tangrams are there?" Now, before we go further we must be sure what is meant by "convex" in this connection. Roughly speaking, we can say that a convex figure is one that does not have any recesses in its outline. For angular figures (like tangram outlines) this means that all the angles are less than 180°; in other words that the corners all stick out instead of in. But the simplest way of seeing the difference between a convex figure and one that is not convex is to imagine a piece of string or an elastic band pulled tight around the figure, as in Nos. 230 and 231, below. If this causes the string to make contact with the figure all the way round its edge, then the figure is convex; but if there are gaps

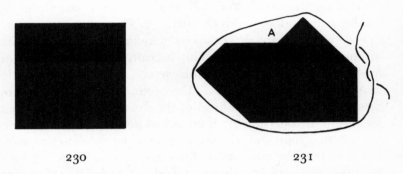

230

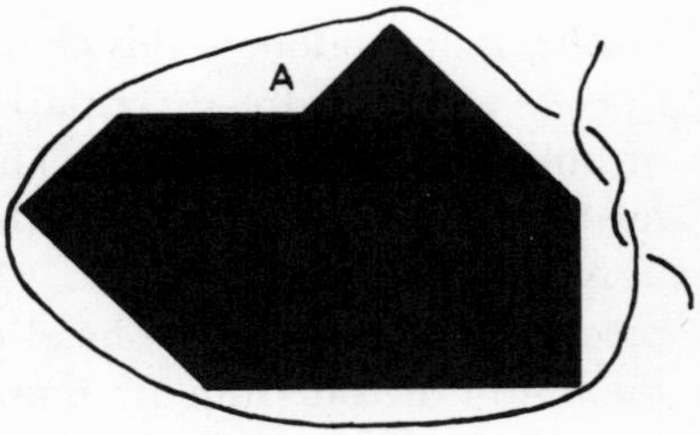

231

between the string and the edge of the figure, as at A in Number 231, then the figure is not convex, but concave or recessed, where these gaps occur.

Now that we know exactly what "convex" means in this context we can look again at the question asked by the Chinese mathematicians: "How many convex tangrams are there?" One might well imagine that there would be quite a large number of them, but it turns out that there are only thirteen! In the paper in which this result is proved* the thirteen convex tangrams are not drawn, but are merely listed as follows:

Triangles	1
Four-sided figures	6
Five-sided figures	3
Six-sided figures	3

In this chapter we give all thirteen convex tangrams. Some of them are very easy to construct, but a few are rather tricky. The reader may want to try to construct them all without first looking at the outlines given below (Nos. 232 to

* Published in the *American Mathematical Monthly*. Vol. 49 (1942), p. 596.

243). Or he can examine the outlines and then try to construct the tangrams. The total of thirteen is completed with Number 230, but since this tangram was necessarily given in the introduction, it doesn't really count.

An Unsolved Problem

It is clear that convex tangrams are very special indeed; there are so few of them. Can we think of any other special kinds of tangrams that would be more numerous than the convex ones, and yet not infinite in number? As far as I know, no one has ever done so, but I am going to propose a problem of this kind for the benefit of anyone who may feel inclined to tackle it. First we must specify the sort of tangram that we are going to talk about.

Let us imagine a set of tangram pieces of such a size that the equal sides of the small triangles are 1 inch in length. Then the third side of these triangles will be of approximately 1.414 inches (the square root of 2, to be precise). Now any side of any of the pieces of this set will be one of these lengths, or twice one of these lengths, and we can therefore imagine every side of each of the pieces to be made up of "sections" whose lengths are either 1 inch or 1.414 inches. There will be either one or two sections to each side. In Figure 3 (page 58), which shows the tangram pieces, the ends of the sections are indicated by blobs.

Imagine now a tangram that has been constructed in such a way that wherever two pieces are in contact at all, they are in contact along a whole section of each, so that the ends of these sections coincide. In other words, when two pieces are in contact, the blobs on the two edges will match. This is illustrated by Figure 4. This stipulation on the way in which the pieces are to be placed together considerably restricts the sort of outline that can be produced; on the other hand, large numbers of very interesting tangrams fall into this category. We shall apply one further restriction: namely, that the tangram should be all in one piece. Tangrams which conform to the above restrictions I call "snug" tangrams, because of the close way in which the pieces fit together. All the convex tangrams are snug; so is Number 231, and so are very many other tangrams in this book. Snug tangrams tend to be rather more difficult to reconstruct from their outlines than tangrams that are not snug, since the close fitting reveals less of the way in which the tangrams have been formed.

It makes sense to ask the question, "How many snug tangrams are there?" for it can be shown that snug tangrams, unlike tangrams in general, are limited in number—there is only a finite number of them. But how many, exactly? At the moment, nobody knows, and I recommend the problem of calculating this number to anyone who finds it interesting and who has access to a large electronic computer. It is unlikely that the "snug tangram number" will be found without the use of a computer, for it is almost certainly very large, probably well up into the millions, if not considerably larger.

(A preliminary investigation of the problem indicates that it is too complex for the computer to which I have access, but that the problem of programming a computer to find the number of snug tangrams would be a fascinating one—a puzzle that out-tangrams the tangram! Unfortunately, it is a puzzle not available to everyone.)

232

233

234

235

236

237

238

239

240

241

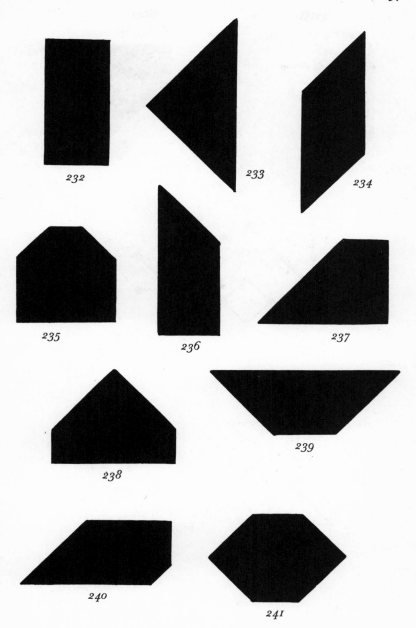

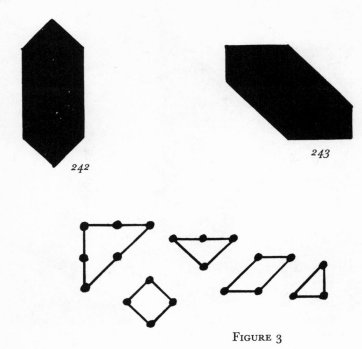

242

243

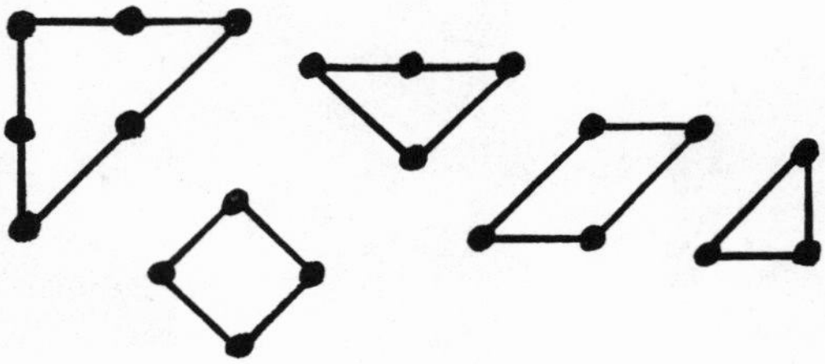

FIGURE 3

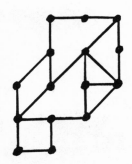

FIGURE 4

8. *Chinoiserie*

As we remarked in the Introduction, the first books of tangrams available in the West were exact reprints of Chinese books on the subject. Some of the objects depicted in these books—ships, birds, and so on—were familiar to Western readers, but the majority of the outlines represented things essentially Chinese, having no Western counterparts. To the editors in England and America these outlines were just so many arbitrary geometrical shapes, and they were treated as such. Yet many of them represent objects of considerable interest, either in themselves or because of their connection with outstanding events in Chinese mythology, literature or history. For this reason we shall look at a few of them in this chapter.

We started this book by forming the letters of the alphabet. The Chinese books did a rather similar thing, but with one important difference. The Chinese schoolboy—pity him!—does not have a paltry twenty-six symbols to learn, as we did, but several thousand characters, many of them very complicated in shape. (To get an idea of their complexity you need only look at our frontispiece, which reproduces the title page of a Chinese tangram book.) Most Chinese characters are far too intricate to be represented in tangram outline, but it is possible to represent some of the simpler ones, and these were not overlooked by the Chinese tangram-puzzlers.

Numbers 244 to 250 give the Chinese characters for "big," "small," "six," "under," "above," "of," and "mountain." (The characters as usually written are: 大, 小, 六, 下, 上, 之 and 山.) The remaining outlines of this chapter are listed below, with comments where necessary.

251. A goldfish. Goldfish came originally from China and are still extremely popular there, as elsewhere.

252. This is called "riding a crane." The ancient

Chinese sages, unlike present-day philosophers, had no desire to go traveling about all over the world, and were quite content to stay in one place to meditate. However, when the need did arise to travel from one place to another, they would do so on the backs of cranes. At least, this is what was popularly supposed; presumably, the cranes of those days were stronger birds than they are now!

The tangram represents one of these immortals mounted on his strange steed. The upper parts of the crane and the rider can be seen, but the nether regions are apparently left to the imagination.

253. A bottle gourd.

254. This is a kind of citrus fruit known as "Buddha's hand."

255. A helmet of a kind once worn by Chinese soldiers.

256. This is entitled "Drinking Alone," and alludes to a very well-known poem by the Tang dynasty poet Li Po, "Drinking Alone by Moonlight." The lonely poet invites the moon and his shadow to join him in his solitary tippling.

257. A clock—probably of Western manufacture.

258. A brazier.

259. A dagger.

260. A pinking knife. (Compare the dressmaker's pinking shears.)

261. A teapot—a familiar enough object to us, too.

262. A mulberry hook—presumably for cutting or trimming mulberry bushes and trees. These would be important to the Chinese, since mulberry leaves constitute the staple of the silkworm's diet.

263. This is a raincoat with a hood. As a tangram outline it shows a clever use of empty space.

264. A soup spoon. Any reader who has had a meal at a Chinese restaurant will know the sort of thing that is meant. It is made of china instead of metal, as our spoons are.

265. Another teapot.

266. A Dragon Boat. This refers to a very old Chinese festival, the Dragon Boat Festival, which was held on the fifth

day of the fifth month. Everyone repaired to the river banks, where boats gaily decked out to resemble floating dragons raced up and down the river. The tangram represents one of these boats.

267. The Three-legged Toad. In China, as elsewhere, there are many stories and legends about the moon, and these stories often center around some interpretation of the markings that can be seen on the moon's disk. According to one such legend the markings represent a three-legged toad that lives in the moon. (Rather like our "Man in the Moon.") The tangram is an attempt (not too successful, I fear) to depict this strange animal.

268. A hexagonal pendant. Another tangram that shows good use of empty space.

269. A carp.

270. A flower bowl.

271. A pair of scissors.

272. A saw. (No handles, apparently!)

273. A clothes post. Two of these were used to support a thin bamboo pole on which clothing was hung to dry. This same tangram is elsewhere described as an incense table.

274. A bell.

275. A feather tube. The ceremonial dress of Chinese generals used to include several very long feathers. (These are often very conspicuous in pictures of scenes from Chinese operas, in which generals frequently appear.) When not in use, these feathers were kept in special tubes in order to prevent their being damaged. In much the same way women on both sides of the Atlantic, around the turn of the century, protected the ostrich feathers they wore in their hats.

276. A lucky charm.

277. A look-out tower.

278. A medicine jar.

279. A rack for a hand mirror.

280. A ladle.

281 and 282. Two baskets.

283. This is a stylized arrow with a flag attached, which the Emperor conferred as a symbol of authority.

284. A "foreign candle"; i.e., one imported from the West.

285. A water pot.

286. A measure.

287. A mallet.

288. A Western knife; i.e., a penknife.

SOME CHINESE CHARACTERS

244

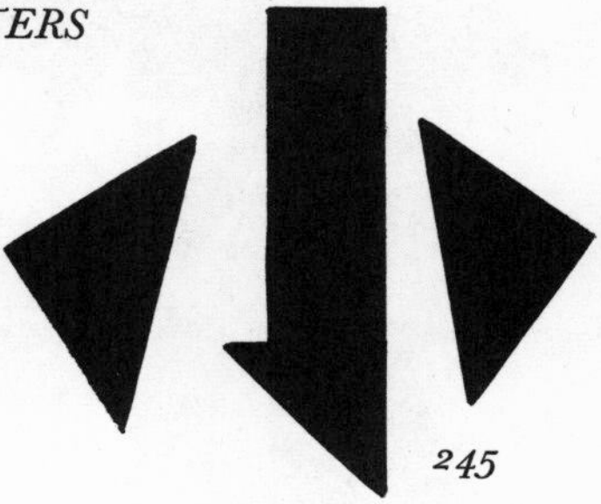

245

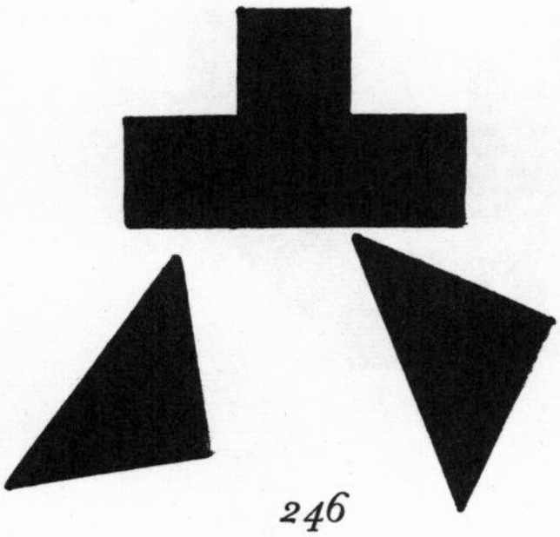

246

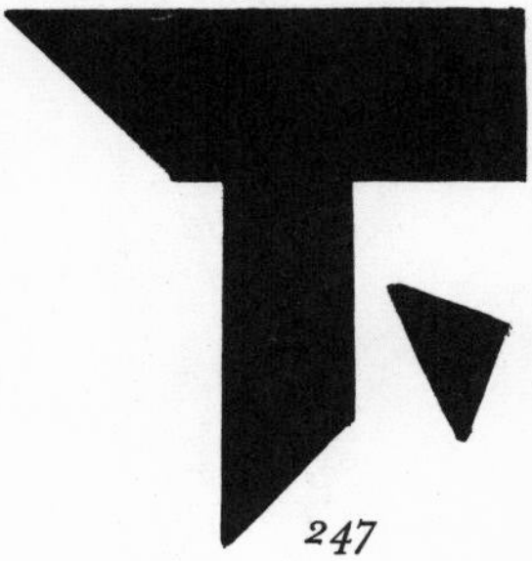

247

248

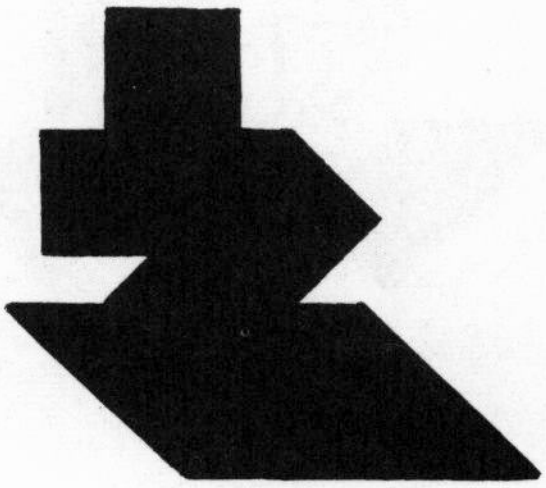

249

250

251 *Goldfish*

252 *Riding a crane*

253 *Bottle gourd*

254 *"Buddha's hand"*

255 *Helmet*

256 *"Drinking alone"*

257 *Clock*

258 *Brazier*

259 Dagger

260 Pinking knife

261 Teapot

262 Mulberry hook

263 Raincoat

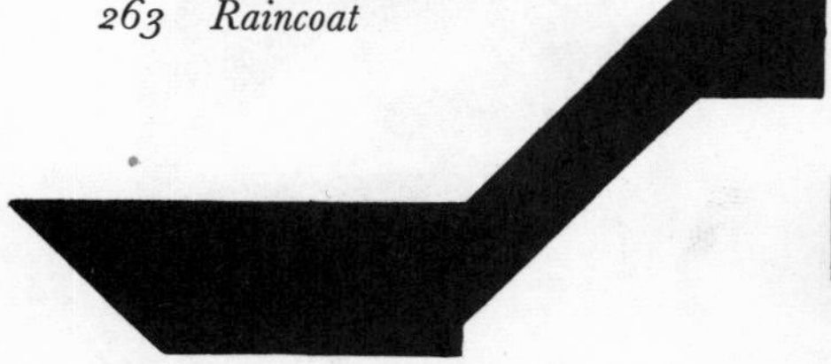

264 Soup spoon

265 Teapot

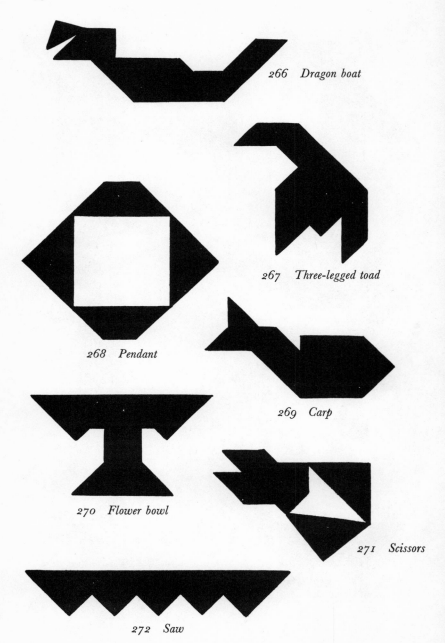

266 Dragon boat

267 Three-legged toad

268 Pendant

269 Carp

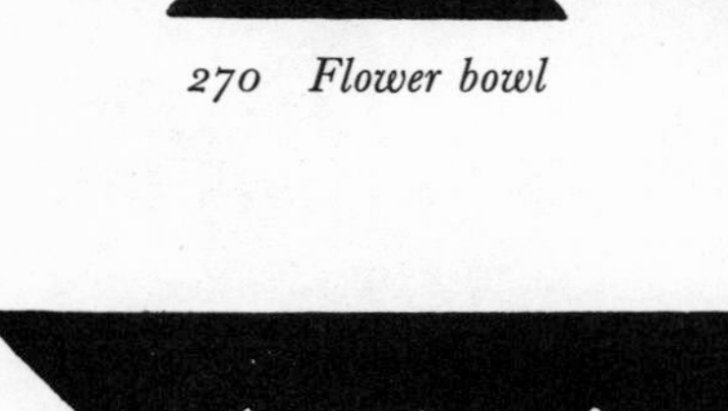

270 Flower bowl

271 Scissors

272 Saw

273 *Clothes post*

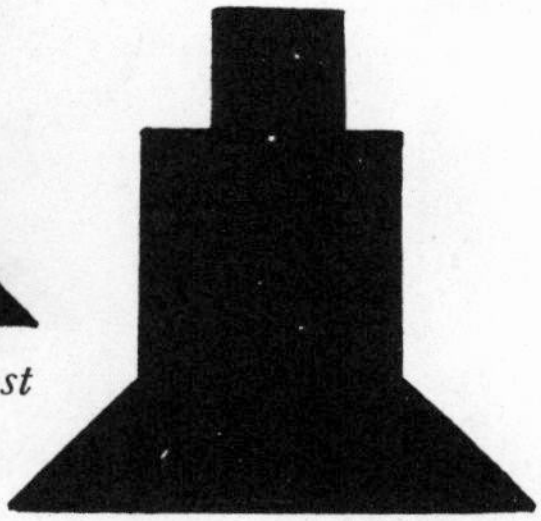

274 *Bell*

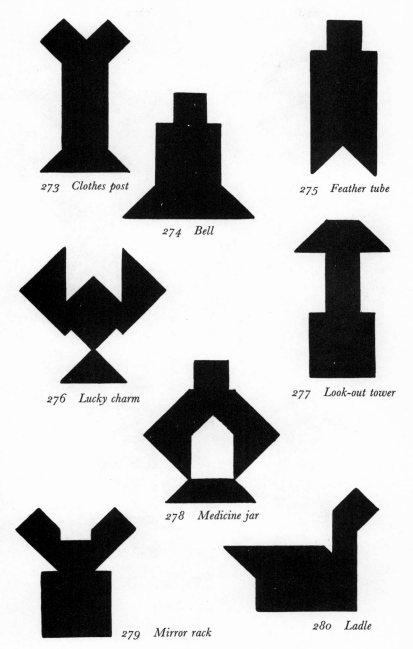

275 *Feather tube*

276 *Lucky charm*

277 *Look-out tower*

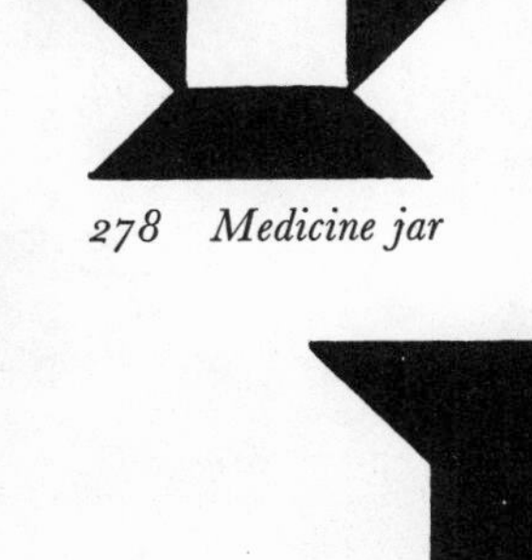

278 *Medicine jar*

279 *Mirror rack*

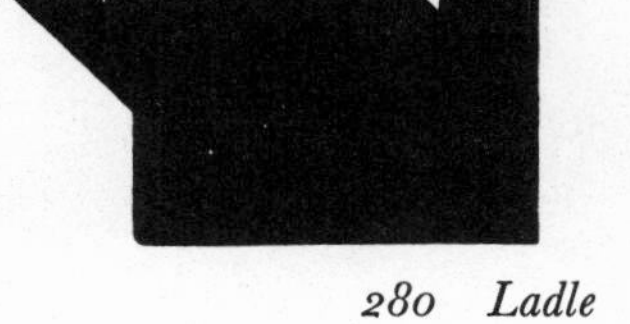

280 *Ladle*

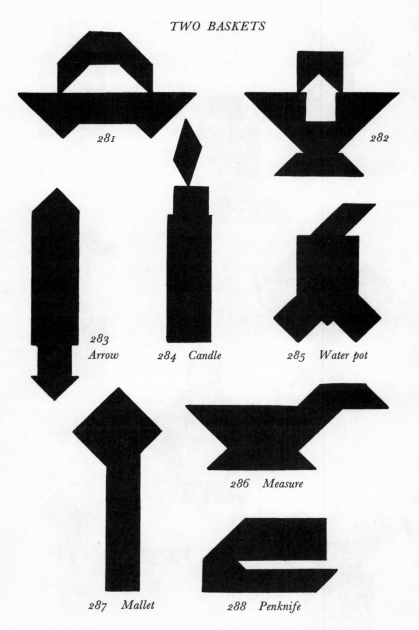

TWO BASKETS

281

282

283
Arrow

284 Candle

285 Water pot

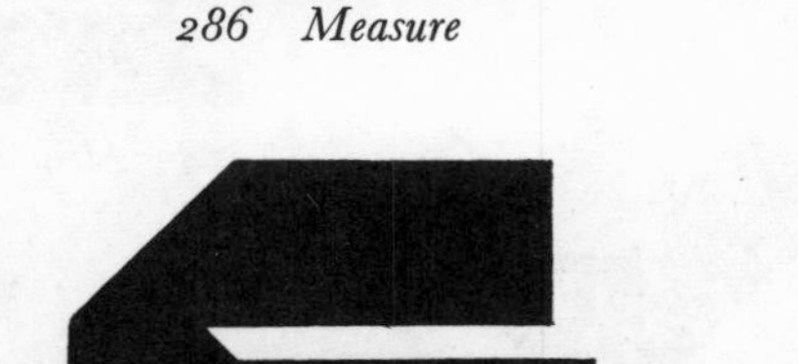

286 Measure

287 Mallet

288 Penknife

9. *Paradoxes and Illusions*

Both Sam Loyd and H. E. Dudeney discovered that the tangram could be the source of some pretty paradoxes. The Chinese puzzlers, it would appear, never hit on this particular aspect of the puzzle. Indeed, their diagrams are so carelessly given (many of them being quite wrong!) that it would have been very difficult for the readers of these books to appreciate the subtle differences between tangram outlines on which these paradoxes depend.

Let us start with a paradox by Dudeney. In Numbers 289 and 290 we have two stately gentlemen who appear to be identical in all respects except that one has a foot while the other has not. Yet each of these tangrams uses all seven of the pieces. Whence, then, does the second gentleman get his foot?

This paradox is similar to, and just as effective as, a paradox that has long been known to mathematicians. Take a square whose sides are, say, 8 inches in length, and divide it into four parts as shown in Figure 5, below. These four pieces

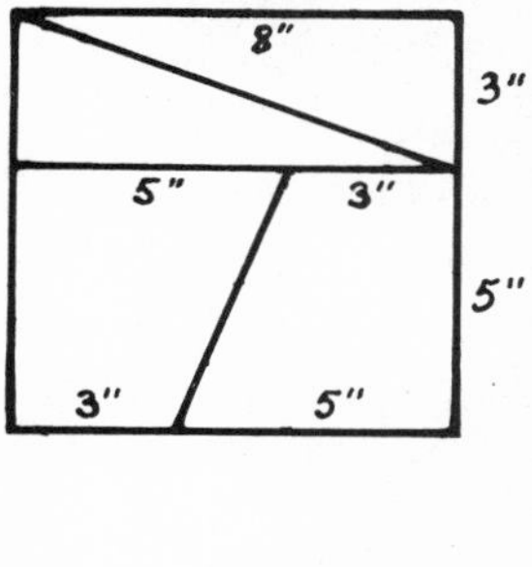

Figure 5

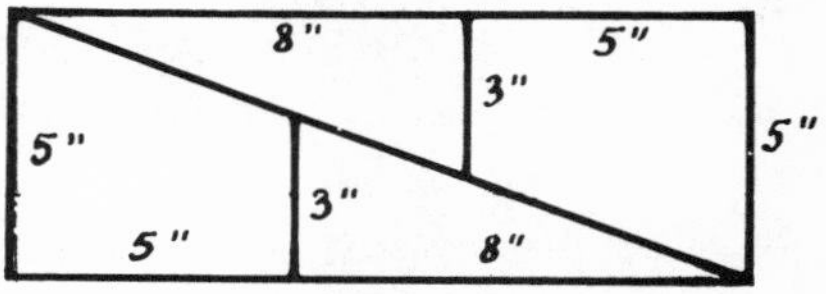

Figure 6

can then be rearranged, as in Figure 6, to form a rectangle measuring 5 inches by 13 inches. At once we see the paradox;

for the area of the original square was $8 \times 8 = 64$ square inches, while the area of the rectangle is $5 \times 13 = 65$ square inches. Where has the extra square inch come from?

If the reader does not believe that the four pieces can be fitted together as shown, I suggest he put it to the test by cutting up a square in the manner indicated, and trying the rearrangement himself. An account of this paradox, together with an explanation of it, and many other paradoxes of a similar kind are to be found in *Mathematics, Magic and Mystery* by Martin Gardener.*

There are many other tangram paradoxes of this kind. Numbers 291 and 292 represent two flatirons, each made with all seven tangram pieces. Yet one of them has an extra knob! Again, in Numbers 293 and 294 we have two snuffboxes, and the knob on the second one is twice the size of the knob on the first!

Then we have four vases (295 to 298), all apparently of the same size. Yet only the last is complete. Number 295 has a chip out of the rim, while 296 and 297 each have a hole in the middle, but holes of different sizes! Tangrams 299 and 300 are not very different in outline, but it requires two entirely different arrangements of the pieces to make them.

The reader who has sorted out the difference between Tangrams 289 and 290 (or who has peeped at the Solutions at the end of the book) should have no difficulty in constructing these tangrams, and explaining the paradoxes that they present.

A different sort of paradox, or perhaps "illusion" would be a better word, was invented by Sam Loyd, and is illustrated by the next eight tangrams. They are those used by Loyd in his book. Number 301 is an ordinary wrench; 302 is a rather more elaborate kind of wrench, with a knob to adjust it. The handle is impossibly short, but let's not be too fussy.

Tangram 303 is a Japanese girl in traditional costume, walking gracefully across the floor; 304 is the same Japanese girl executing a polite bow, displaying the bustle which is a characteristic part of the traditional dress. Tangram 305 is yet

* New York: Dover Publications, Inc.

another baby carriage, and 306 a version with large wheels in front.

The reader should try to construct these six tangrams before reading any further.

After the reader has constructed these tangrams he should answer (honestly) the following question: Did you notice that Tangrams 302, 304 and 306 are identical, except for the position in which they are drawn?

A similar illusion is presented in Tangrams 307 and 308, showing a man pushing a wheelbarrow. One could very easily overlook the fact that these two tangrams are exactly the same!

We return to the original type of paradox in Tangrams 309 and 310. One of these pigs has lost his tail. Where did it go?

Finally, from the tangram pieces, which were originally cut from a complete square, we are asked to make a square with a square corner removed from it. The side of the piece removed is one-third the length of the side of the big square. Despite the apparent paradox, this can be done. What is more, the seven tangram pieces can be divided into three groups in such a way that the transition from the complete square to the square-with-a-bite-out-of-one-corner can be made simply by moving these groups *en bloc*. Solve this tangram puzzle and you will have solved the following dissection problem: With two straight cuts, divide a square into three parts which can be rearranged to form a square with a square portion (of dimensions given above) missing from one corner. These two cuts are indicated by extra heavy lines in the Solutions section at the end of this book.

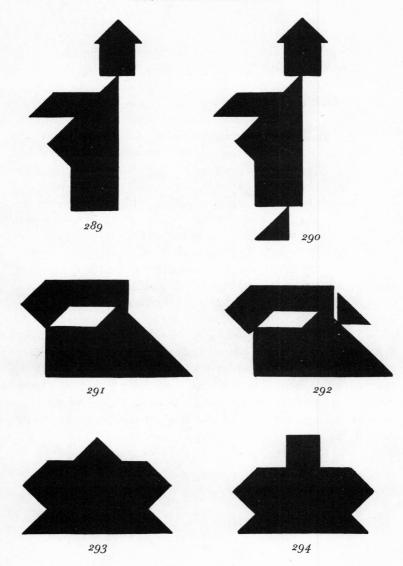

289

290

291

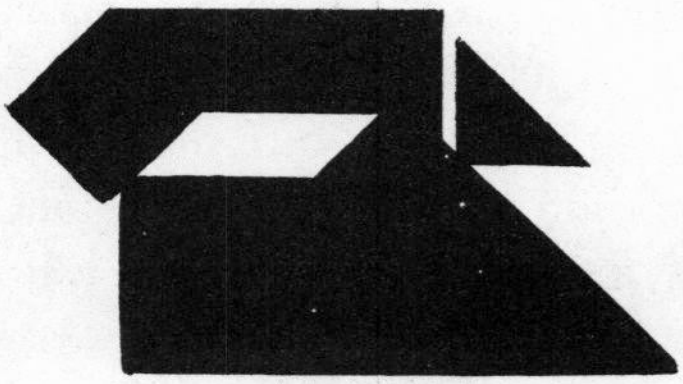

292

293

294

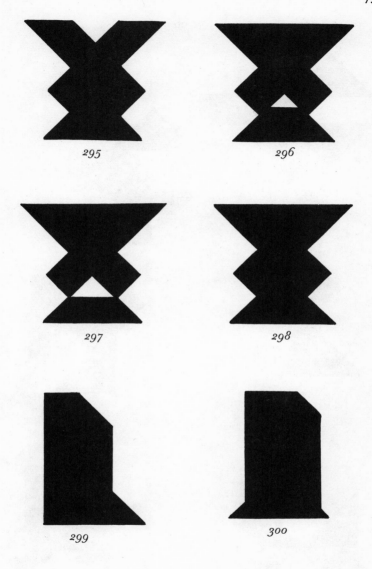

295

296

297

298

299

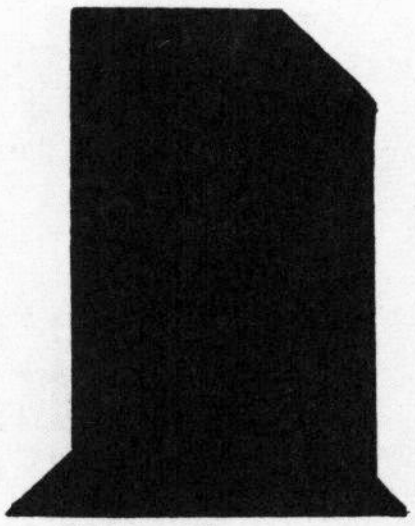
300

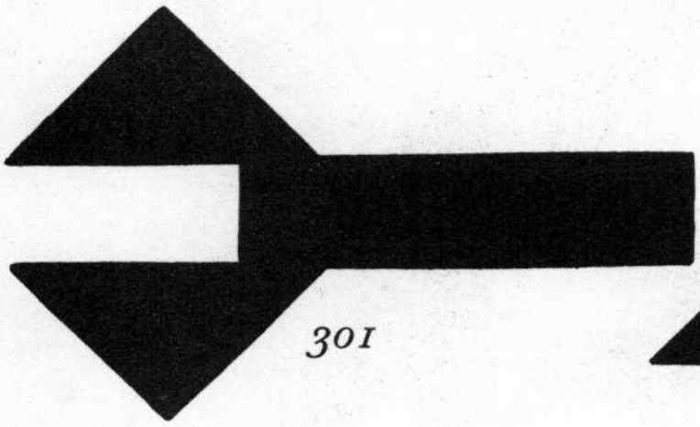

301

302

303

304

305

306

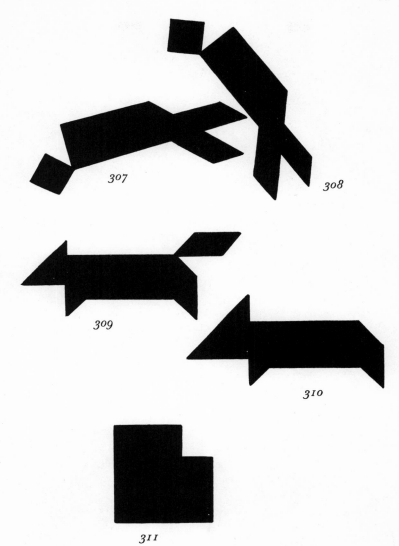

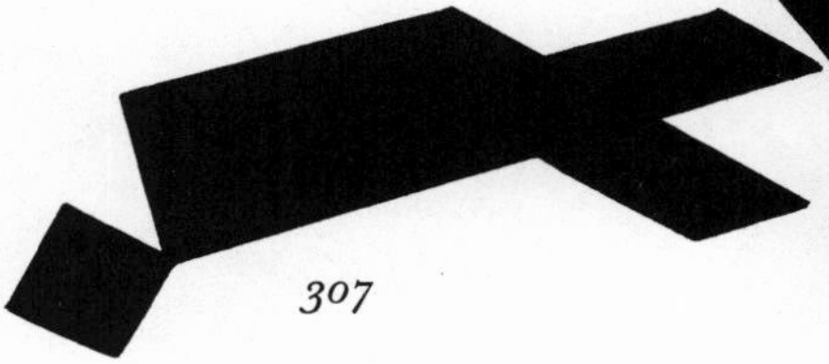

307

308

309

310

311

10. Double Tangrams

TANGRAMS are amazingly varied and numerous, but there are, of course, limits to what can be done with only seven pieces. It is natural, therefore, to try the effect of using two or more sets of tangrams in an endeavor to achieve more intricate results. Dudeney's pictures (Chapter 6), even though they use several tangram sets, are nevertheless in the true tangram tradition, since each element in the picture is a single complete tangram. In this chapter we shall look at some "double tangrams," made from two sets of tangram pieces, in which the fourteen pieces are used all together, with no separation between the pieces of one set and those of the other.

As the reader will see from the outlines given in this chapter, much more intricate shapes are possible with a double tangram set; yet, on the whole, these double tangrams are less interesting than their single brethren. With fourteen pieces to play around with, one cannot help but feel that it should be possible to arrive at a reasonable likeness to just about anything. Consequently, the sense of achievement that one gets on producing a recognizable cow, sailing boat, human figure, or what have you, from a mere seven pieces, is quite lacking. For all that, the double tangrams are not altogether without interest, and certainly deserve mention. In the following pages we give a selection of double tangram outlines.

Numbers 312 to 317 are self-explanatory (I hope). Number 318 is Cinderella's coach, which we promised in Chapter 6. (If you think it still looks like a baby carriage, I can only say, "Go and make a better one, then!") Then we have a space rocket and a television set (We're really up to date now!), followed by a locomotive. Finally, a microscope (322), a telescope (323), and that most fiendish of all inventions of the devil, an alarm clock!

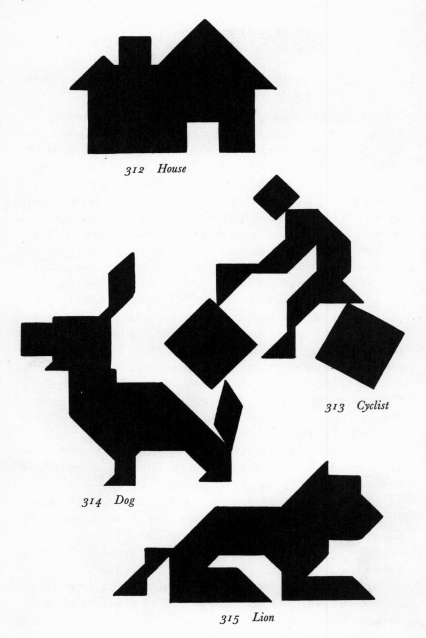

312 House

313 Cyclist

314 Dog

315 Lion

316 Car (ancient vintage)

317 Sports car

318 Coach

319 Rocket

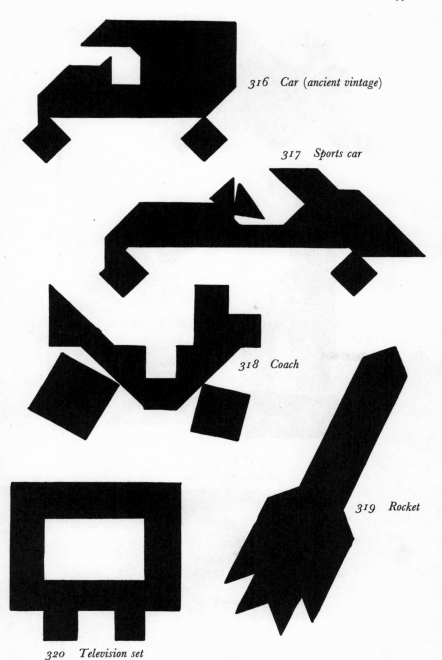

320 Television set

321 Locomotive

322 Microscope

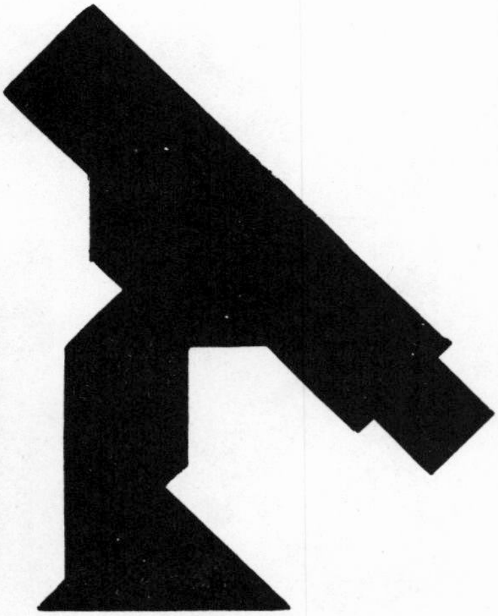

323 Telescope

324 Alarm clock

11. *The Fifteen-Piece Puzzle*

IN 1962, while searching in the British Museum for further information on tangrams I came across two volumes printed in China in the latter half of the nineteenth century dealing with a puzzle which will be of interest to those who find the tangram interesting. These books were written by one T'ung Hsieh-keng, and the Chinese name for the puzzle is literally "Increase-knowledge puzzle"; we shall call it the fifteen-piece puzzle. As this name suggests, fifteen pieces are used, and they are cut from a square in accordance with the following figure:

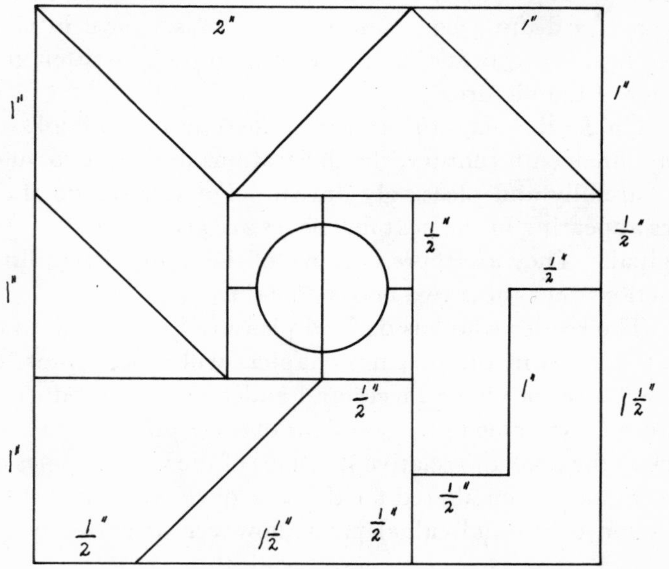

FIGURE 7

Suggested measurements are given, though the square can be of any size. With these measurements the diameter of the central circle is $\frac{2}{3}$ inch. As with the tangram pieces, all the angles are either 45° or 90° or 135°.

What was said in the last chapter about the lesser interest of the double tangrams, compared with single tangrams, will apply equally well here; with fifteen pieces one ought to be able to make a reasonable outline of almost anything. The fifteen-piece puzzle has some added points of interest, however, in that the pieces are more varied in shape; there are even some curved edges.

The two Chinese volumes contain a large number of pictures made up from the fifteen pieces. Many of them are straightforward outlines of single objects, not so very different from the tangrams or double tangrams given in this book; but many of them are complete pictures, very much in the traditional style of Chinese painting, and with a typically Chinese choice of subject matter—a pavilion in the moonlight, a man asleep in a drifting boat, and so on. As is typical in Chinese paintings, an appropriate poetic quotation is written in one corner of the picture.

Unlike the diagrams in the Chinese tangram books of the early nineteenth century, the illustrations in these two volumes are carefully and pleasingly drawn. For this reason the outlines appearing in the next few pages are given exactly as in the original. They are representative of the many delightful constructions that these two books contain.

The reader who has derived pleasure from trying to make the tangrams in the first ten chapters will surely want to cut himself a set of fifteen pieces and endeavor to reproduce these pictures. In some of the solutions of these puzzles given at the back of the book the relative positions of the various parts of the picture have been altered for the sake of convenience; the construction of the individual parts, however, is correctly given.

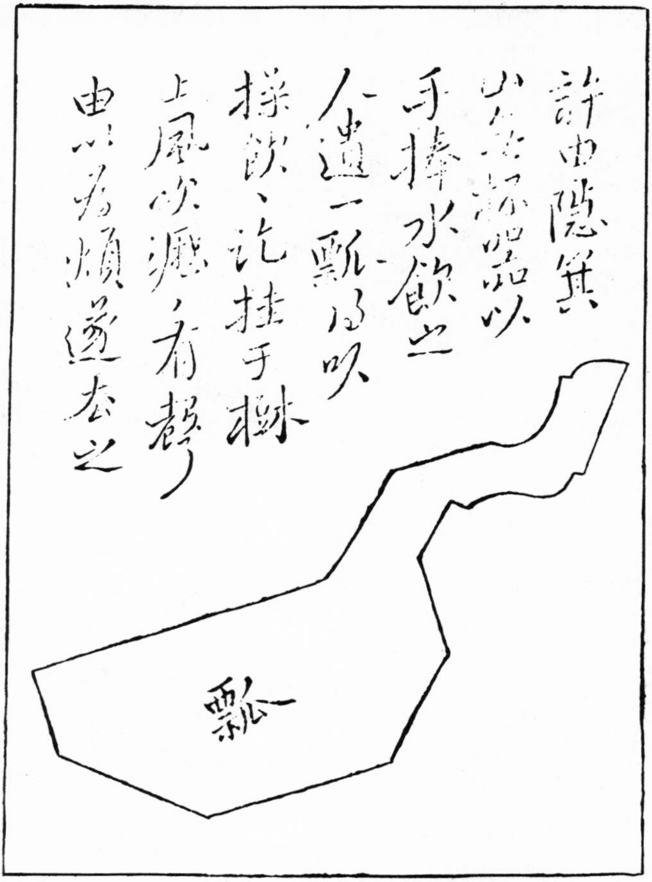
許由隱箕
山无杯器以
手捧水飲之
人遺一瓢得以
操飲、訖挂于樹
上風吹瀝〻有聲
由以為煩遂去之
一瓢

瓠匜其腹
錢大其口
斟之酌之
呼為陶友

古陶器

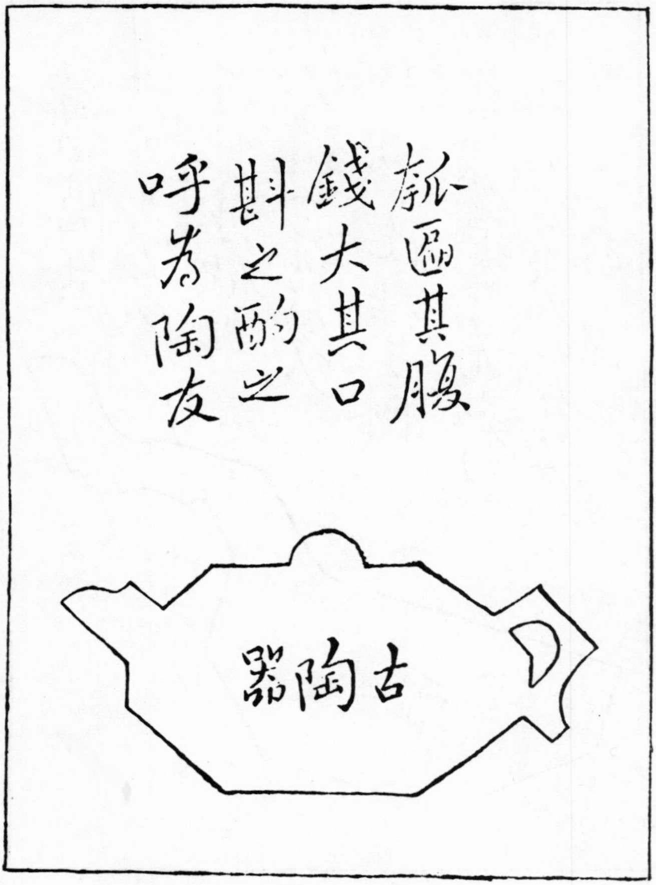

不炊而沸
不汲自盈
絪縕之氣
自然所生

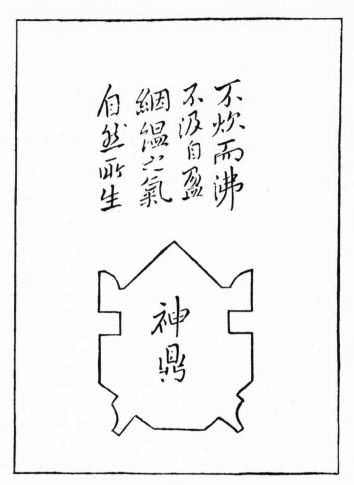

327

鼎三足兩耳和五味之珤器也

328

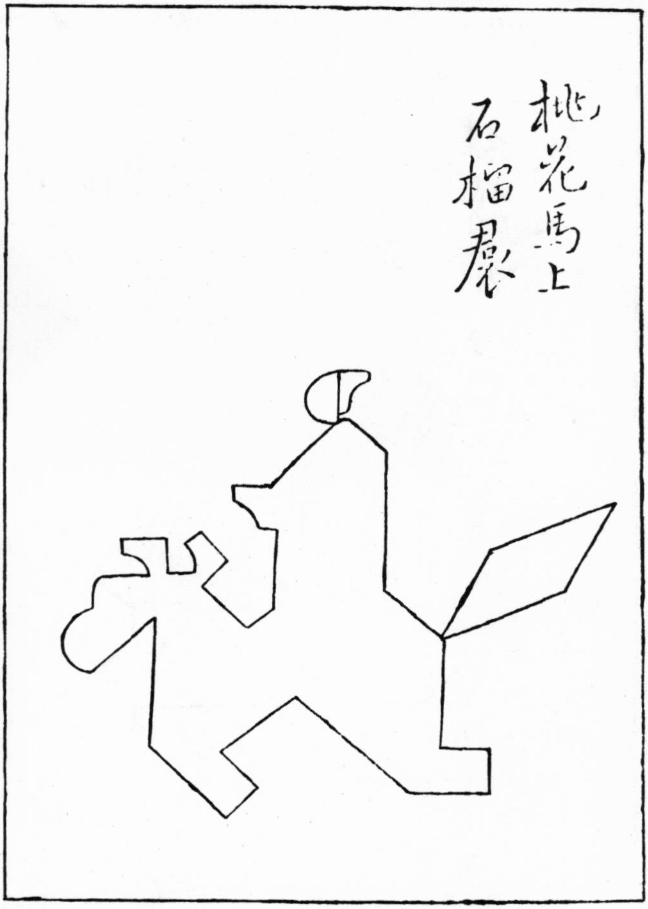

329

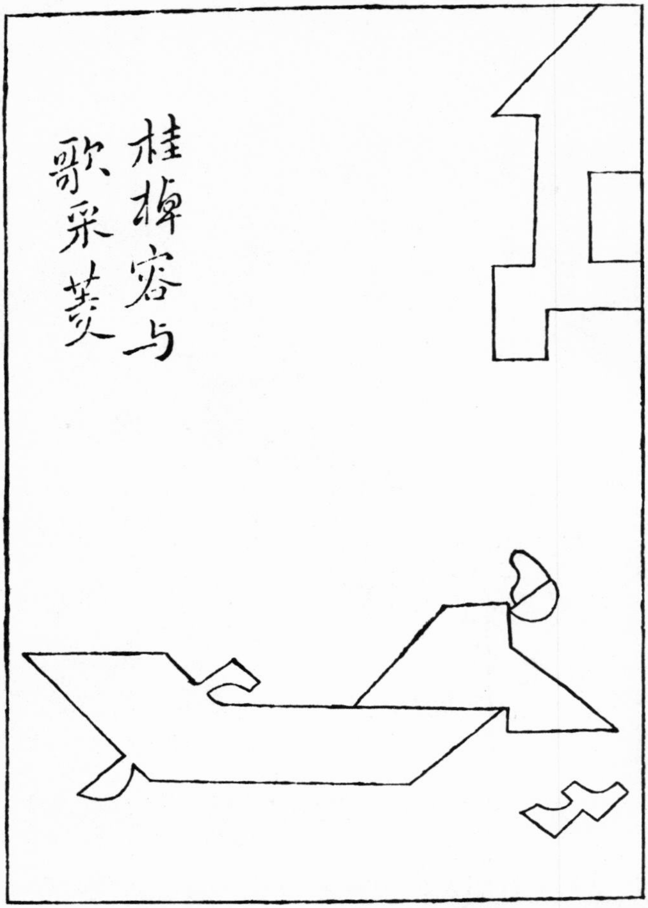

330

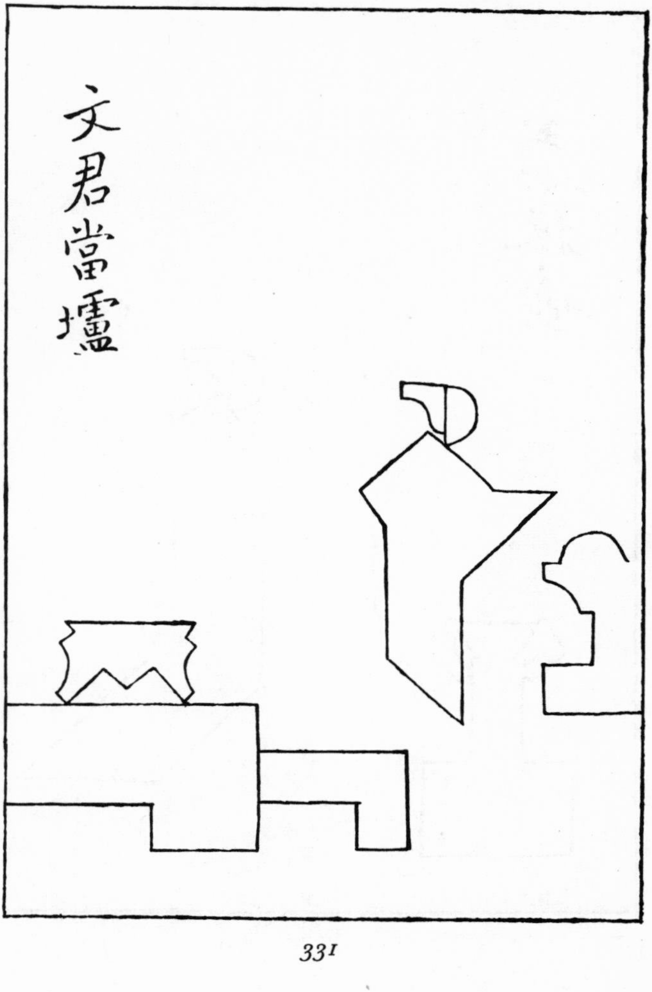

331

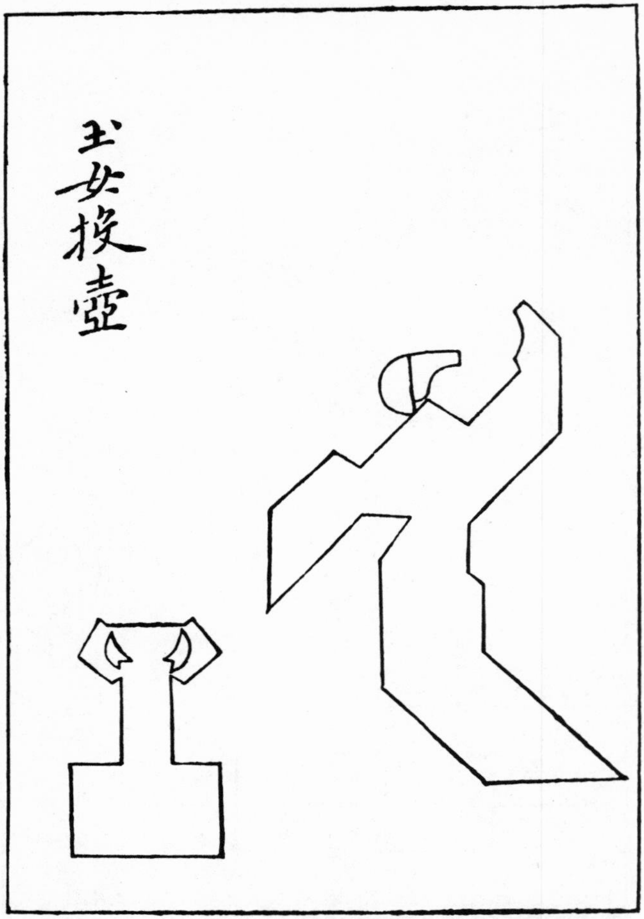
玉女投壺

十扣柴扉九
不开一枝红
杏出墙来

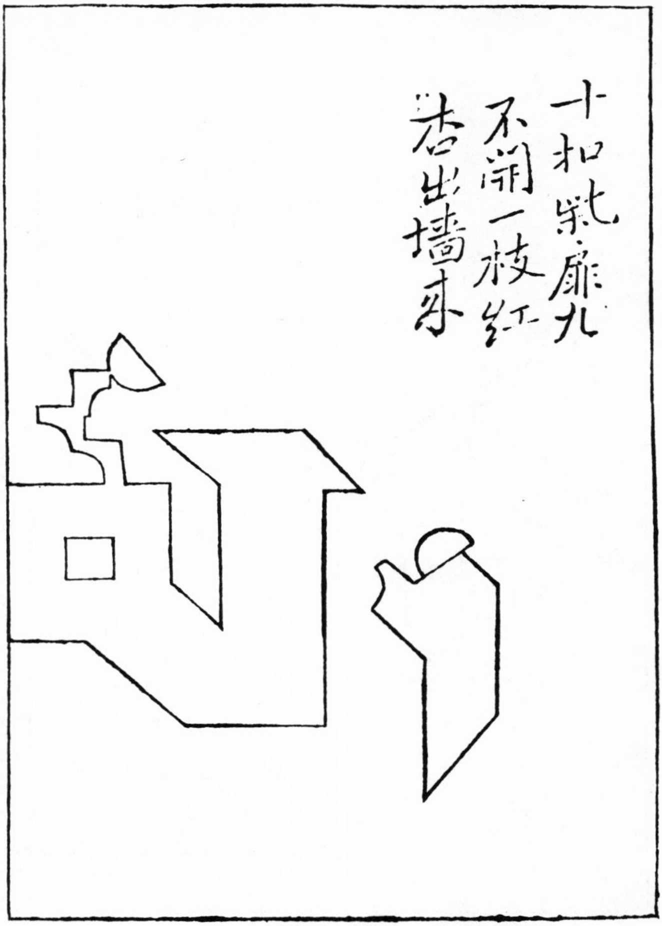

333

姑蘇城外寒山寺夜半鐘聲到客船

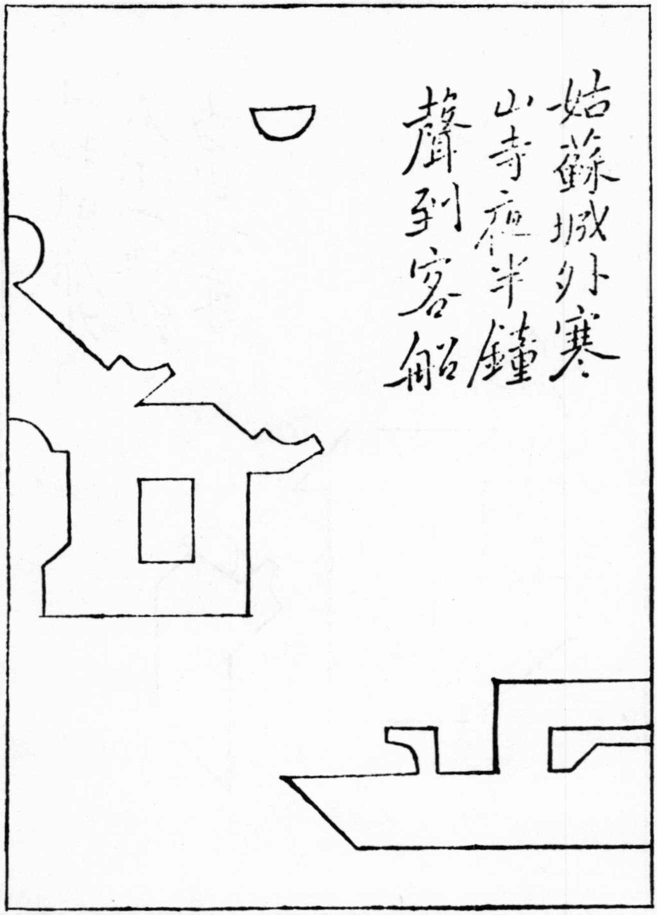

334

Solutions

1

2

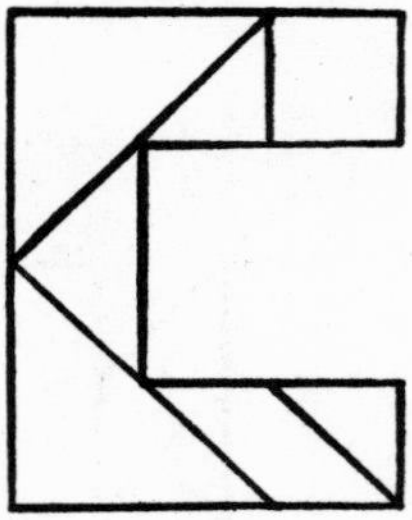

3

4

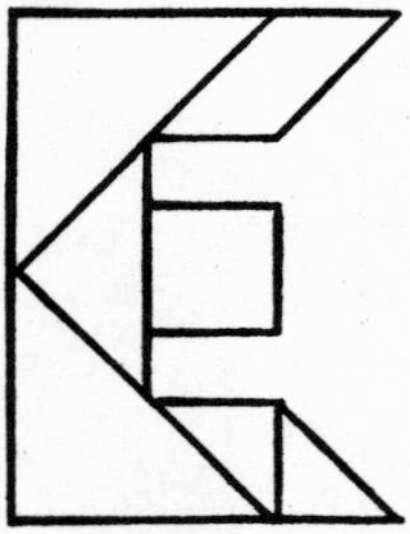

5

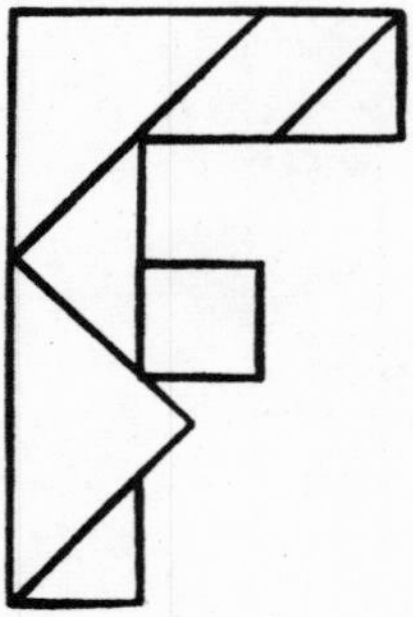

6

7

8

9

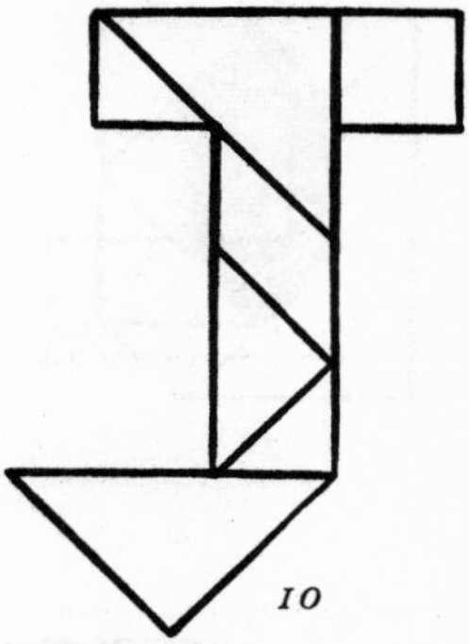

10

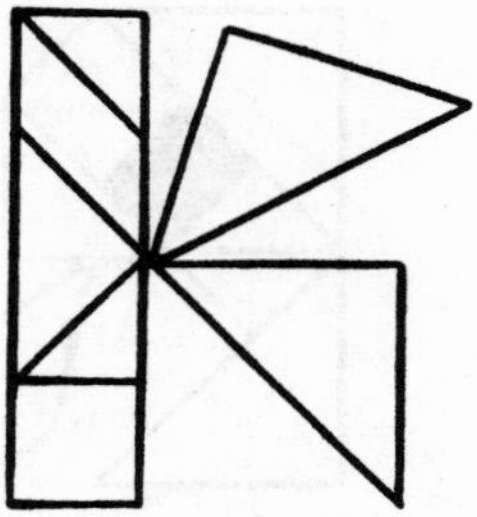

11

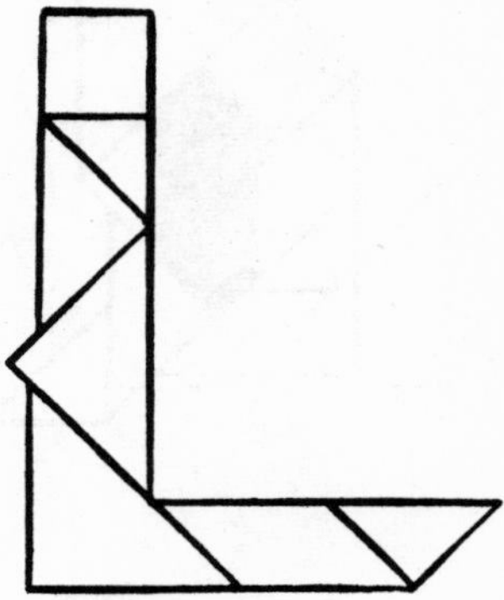

12

13

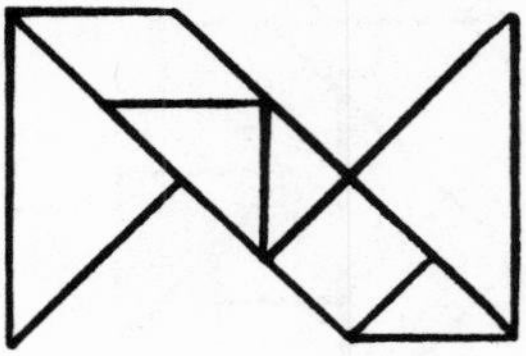

14

15

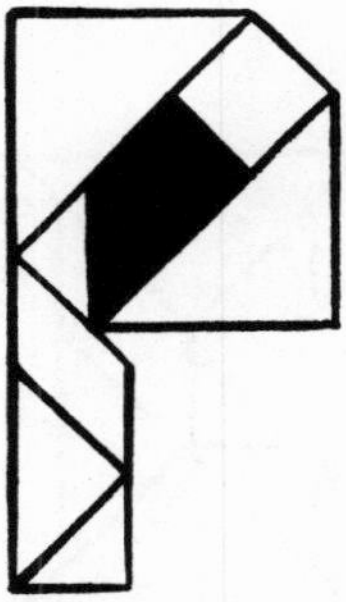

16

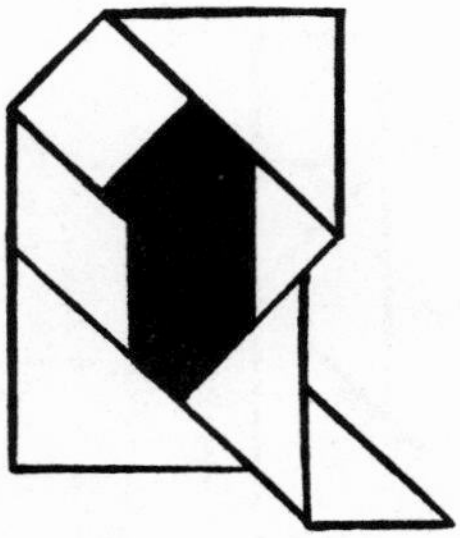

17

18

19

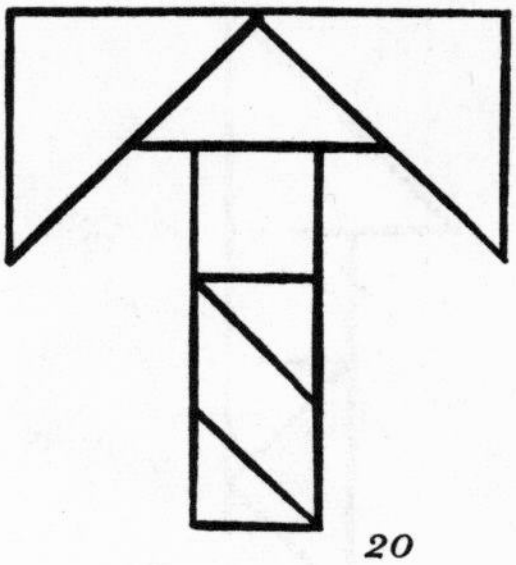

20

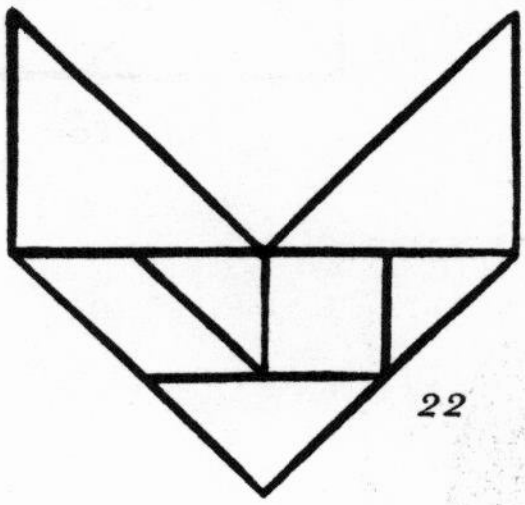

22

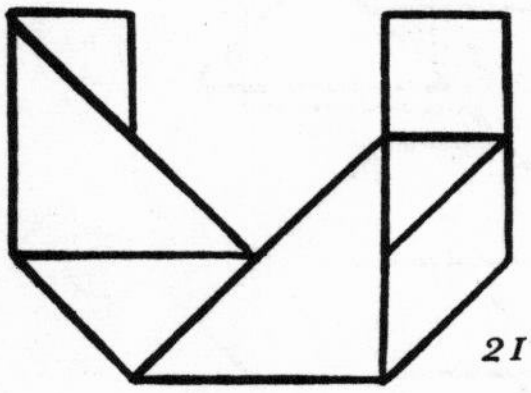

21

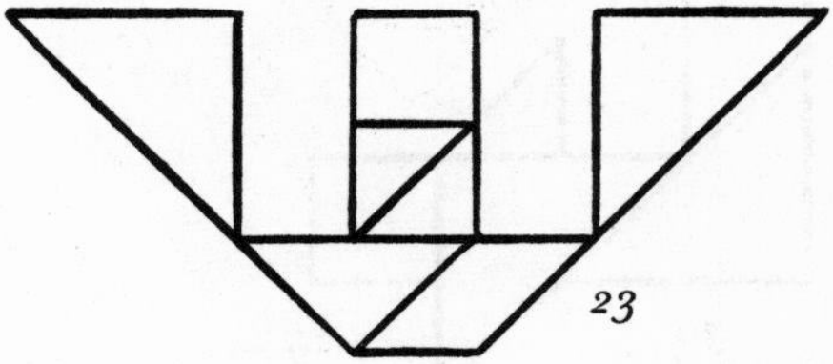

23

24

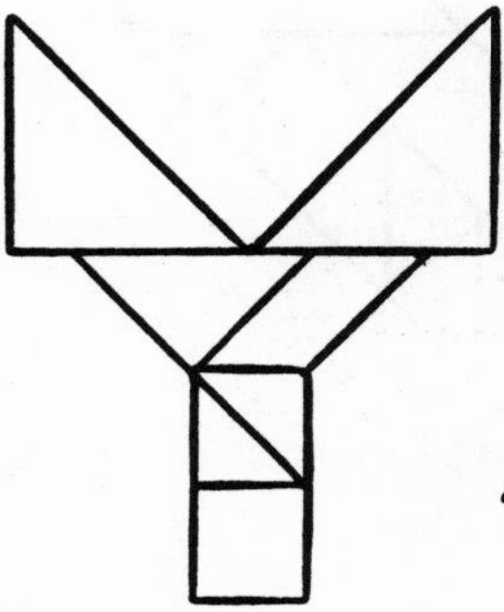

25

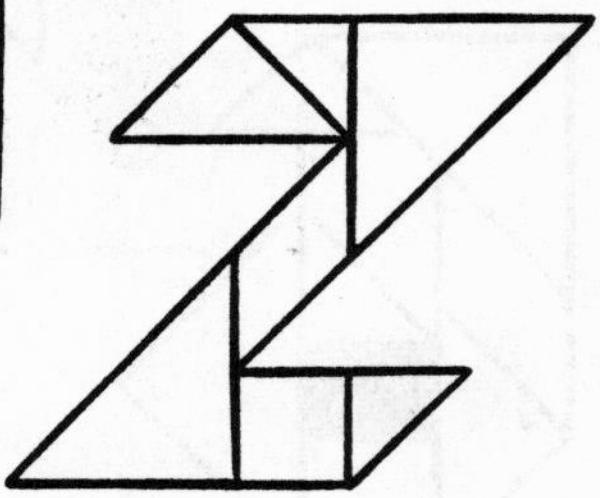

26

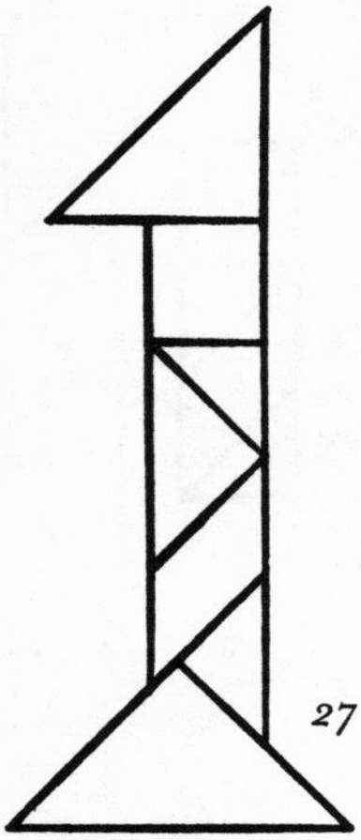

27

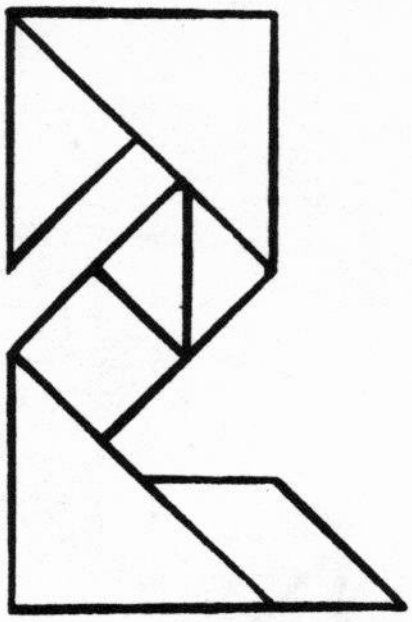

28

29

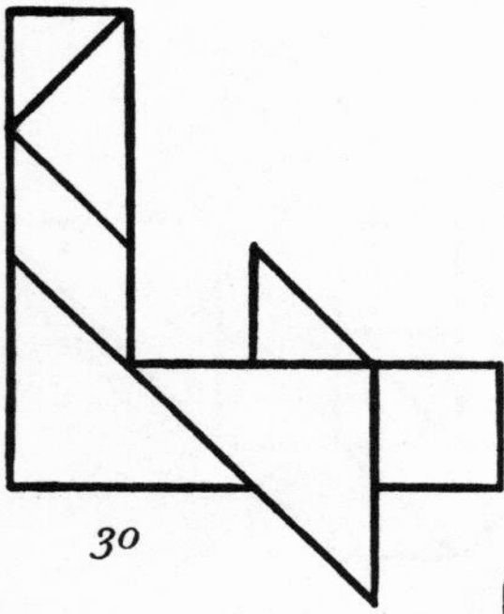

30

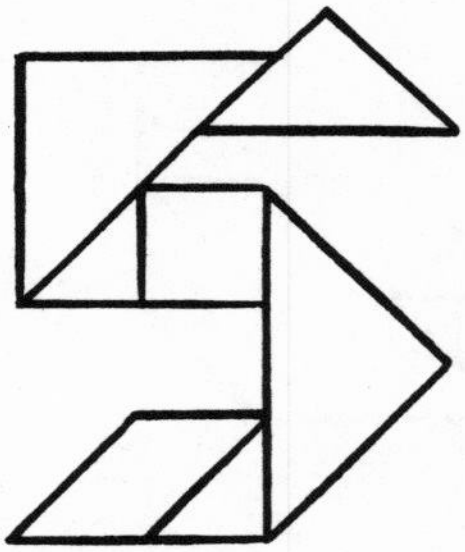

31

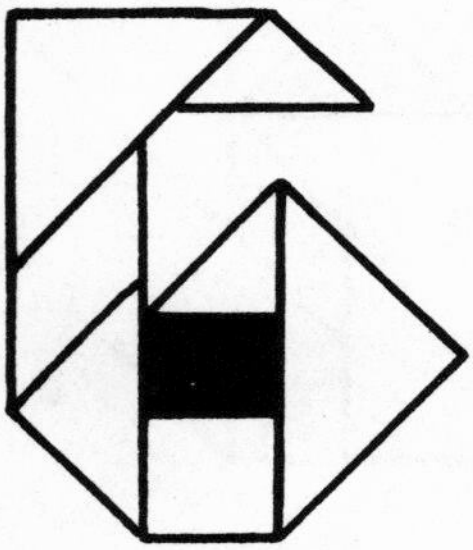

32

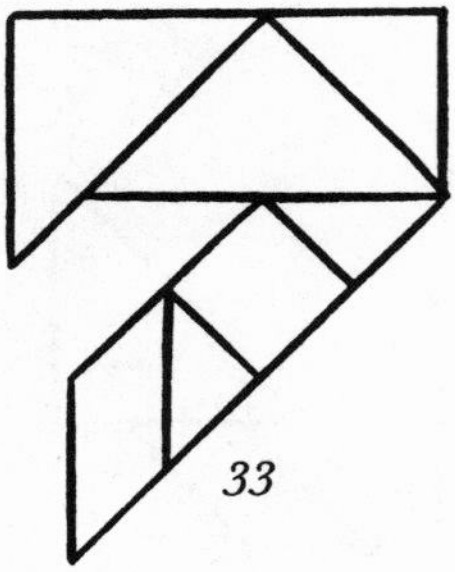

33

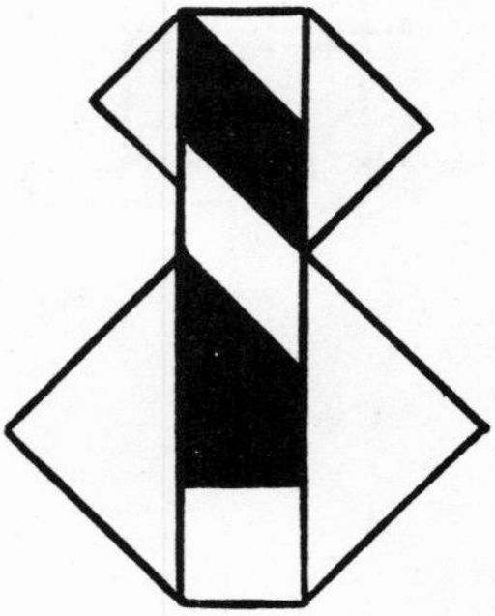

34

THE TANGRAM ZOO

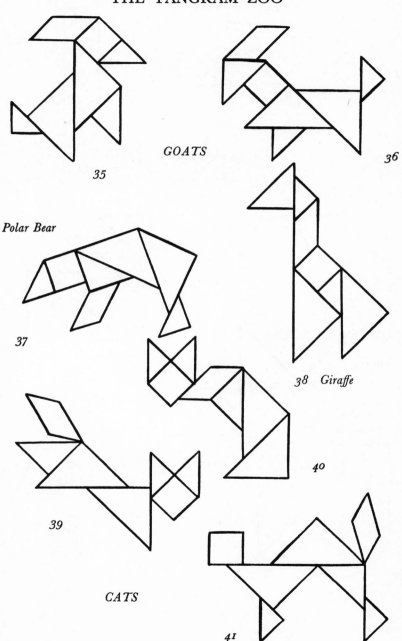

GOATS

35

36

Polar Bear

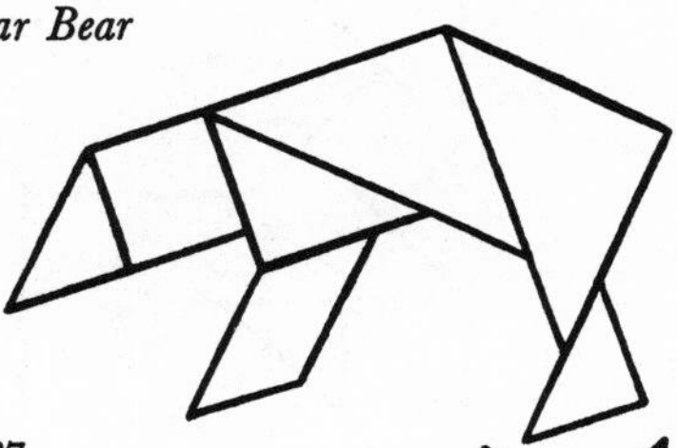

37

38 *Giraffe*

40

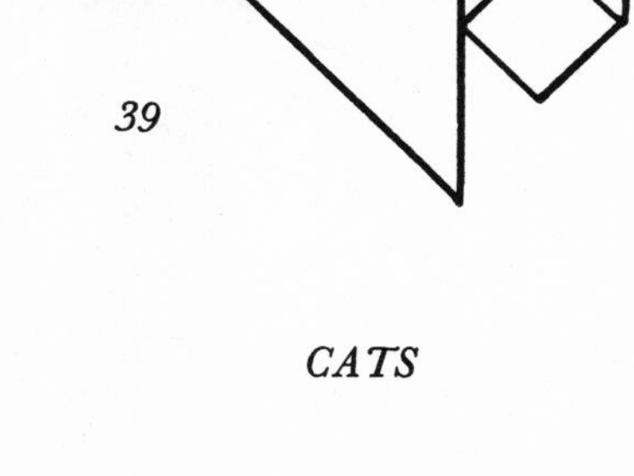

39

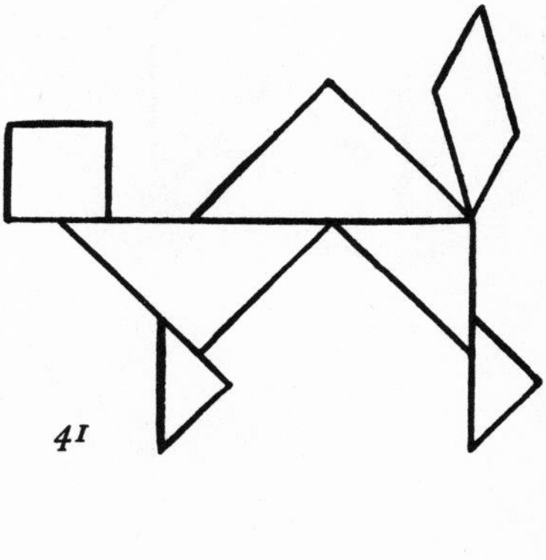

CATS

41

CATS AND DOGS

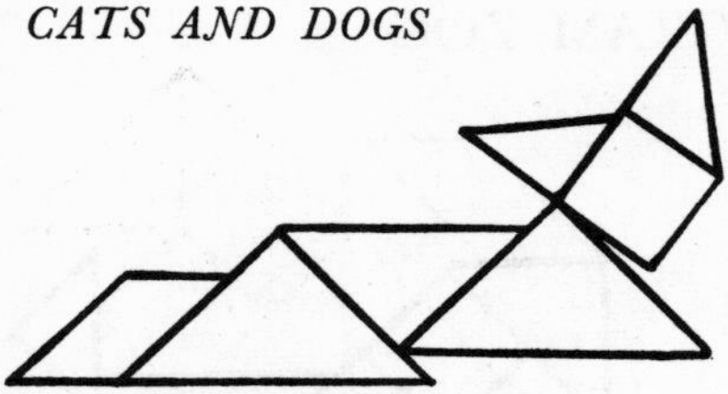

42

43

44

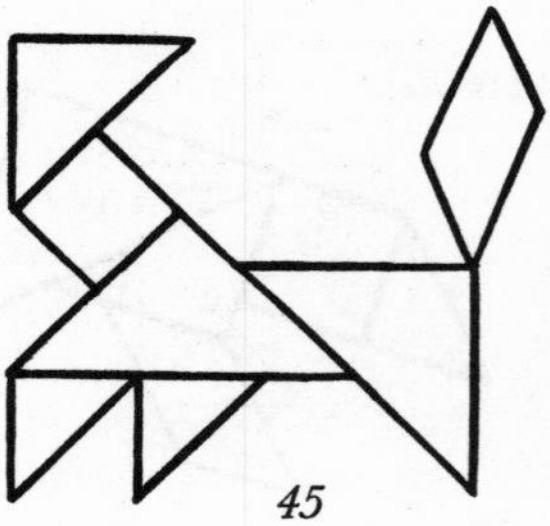

45

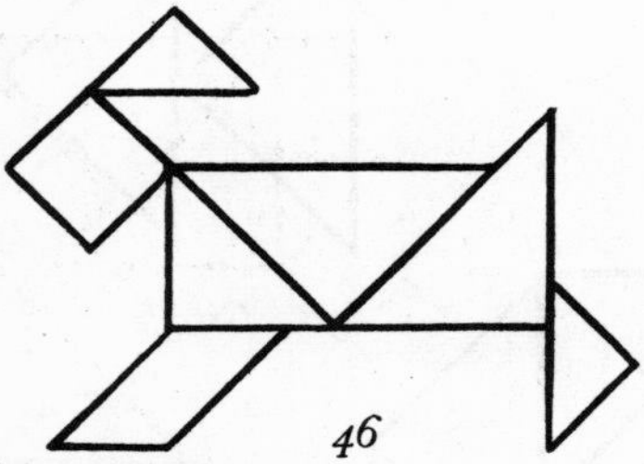

46

47 Camel

48 Squirrel

BIRDS

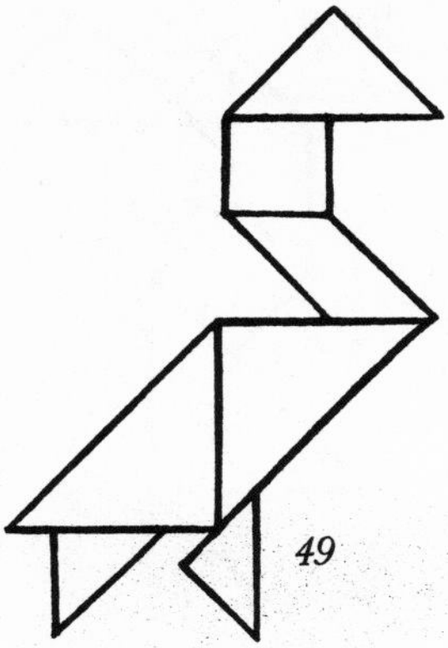

49

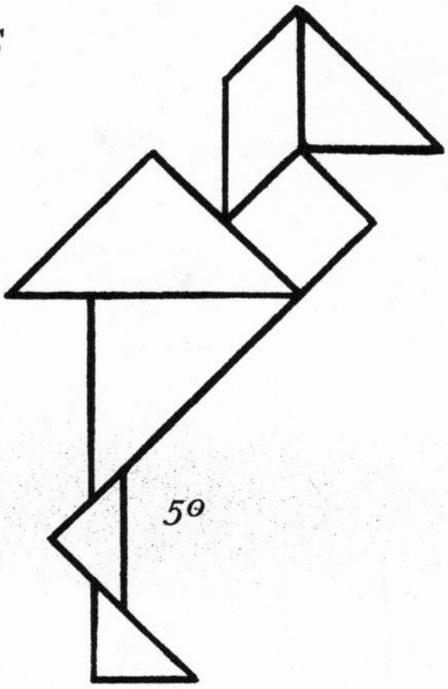

50

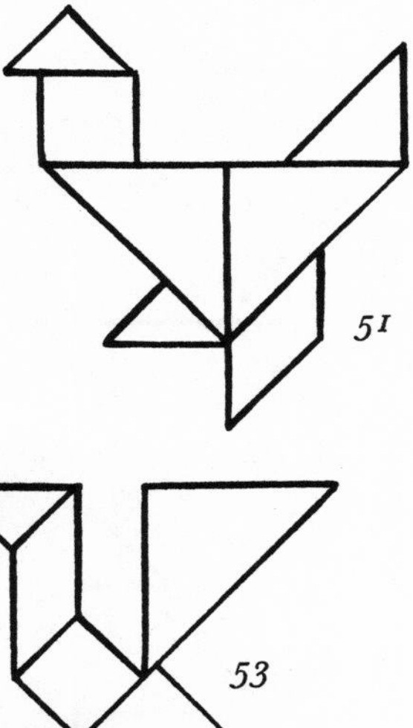

51

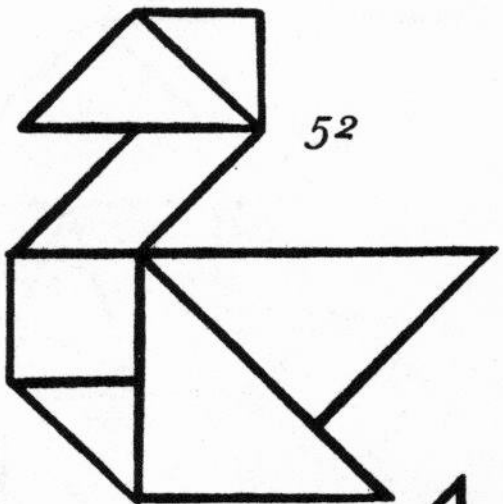

52

53

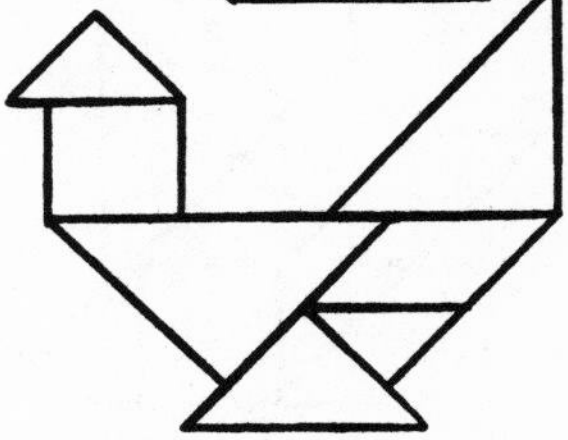

54

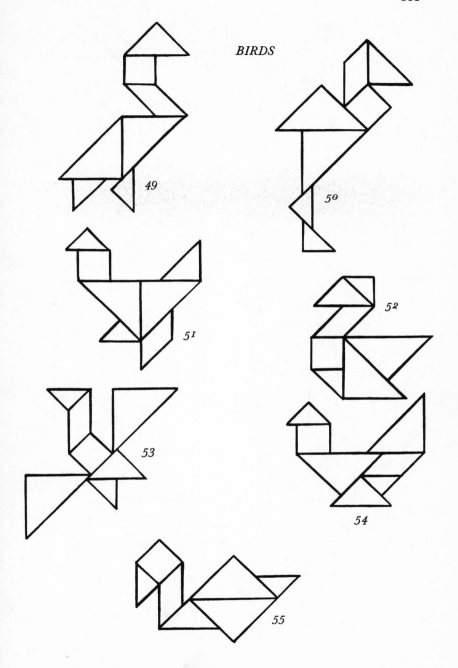

55

MORE BIRDS

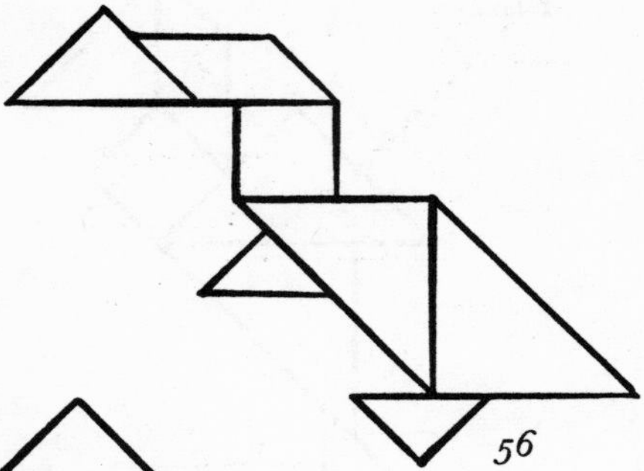

56

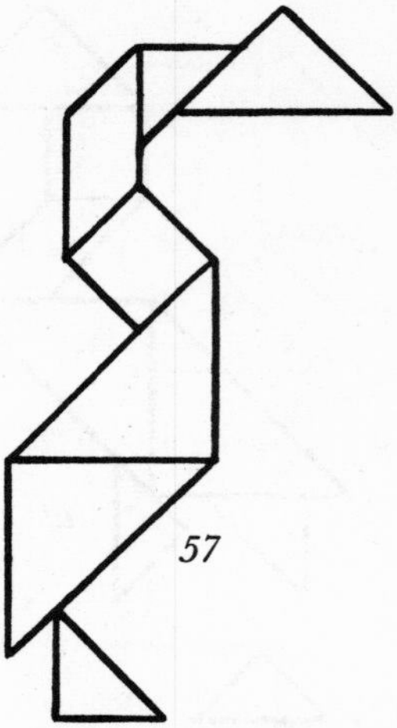

57

58

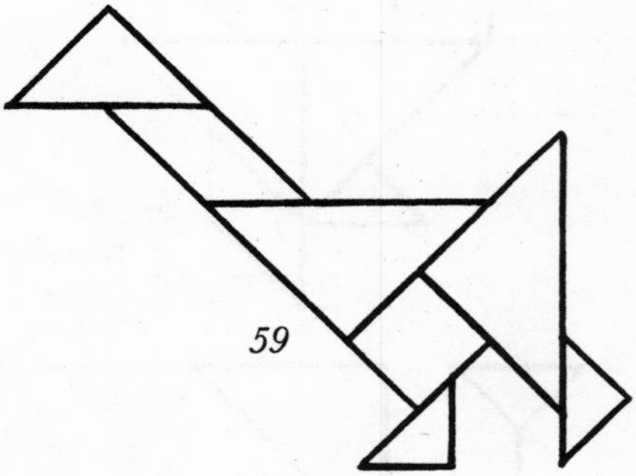

59

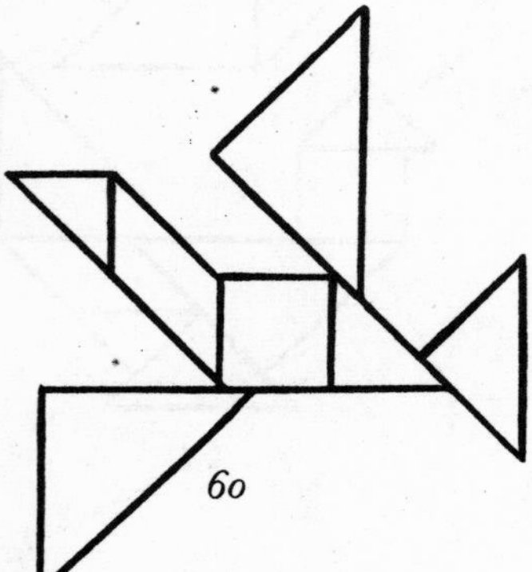

60

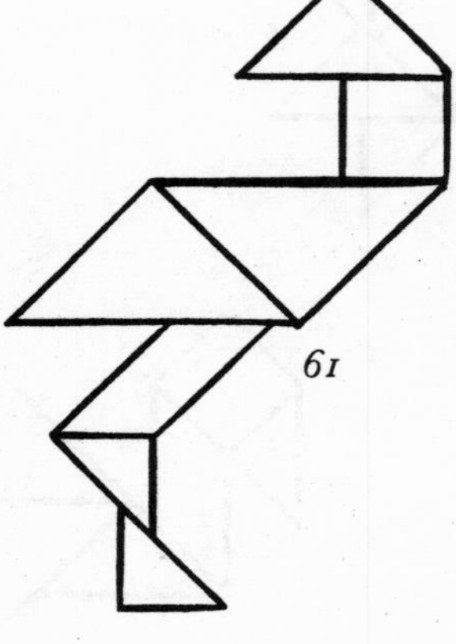

61

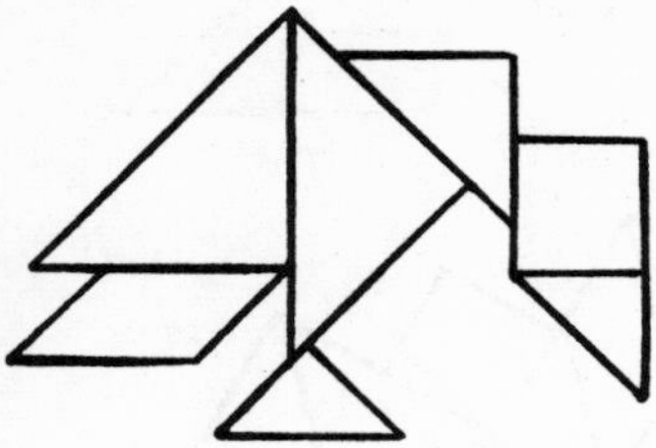

62

63

VULTURES

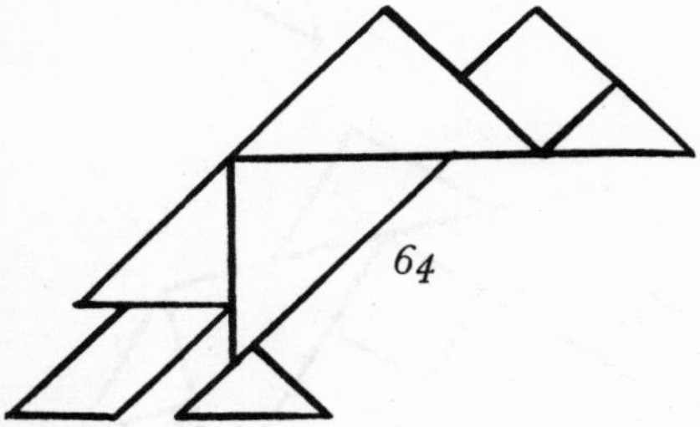

64

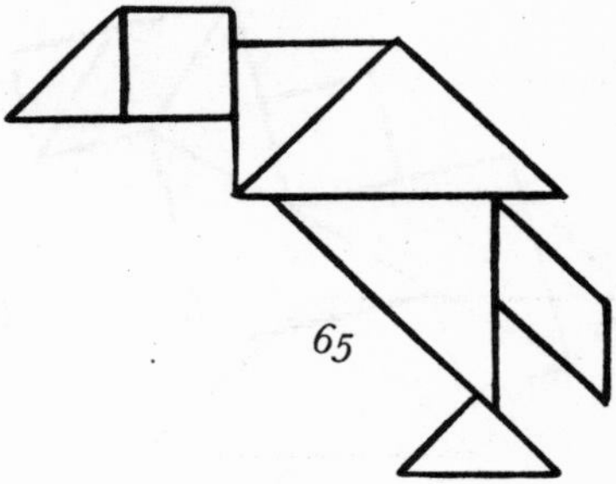

65

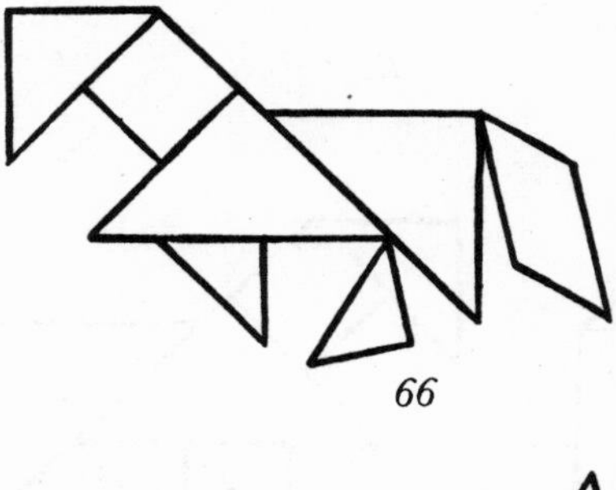

66

67

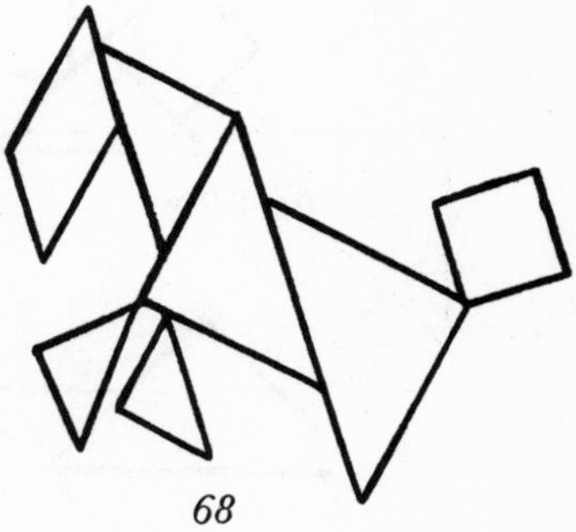

68

HORSES

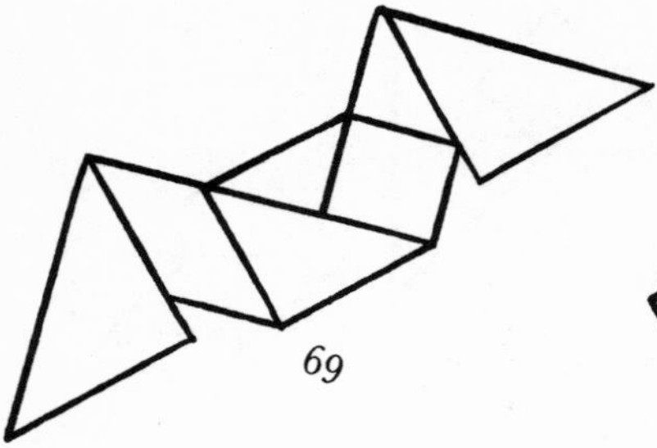

69

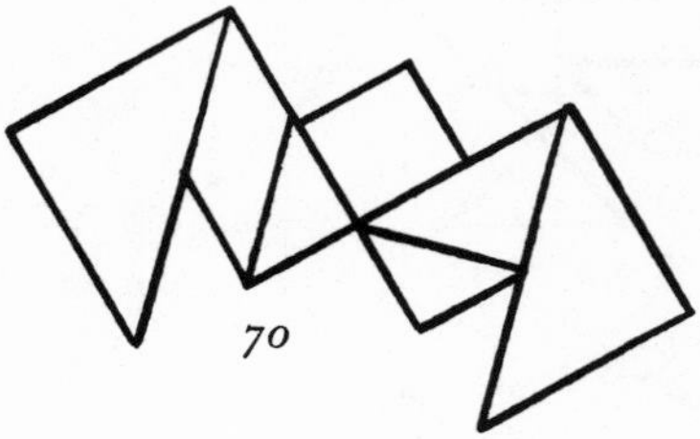

70

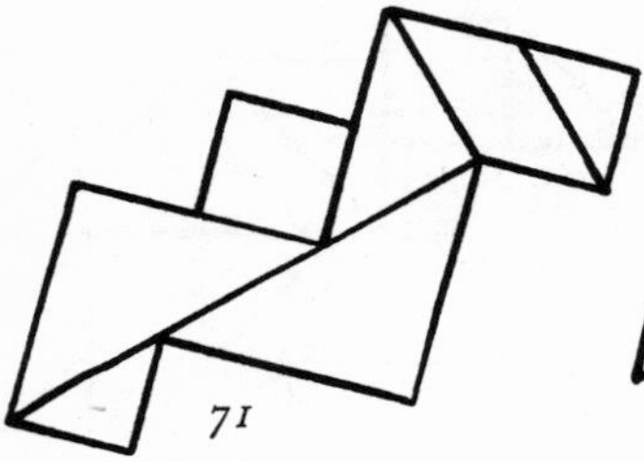

71

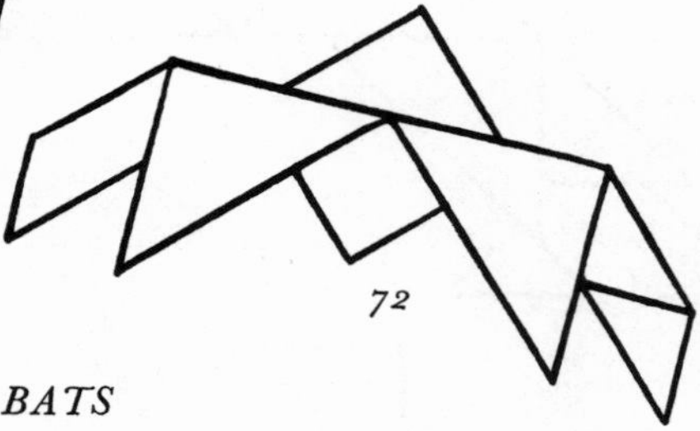

72

BATS

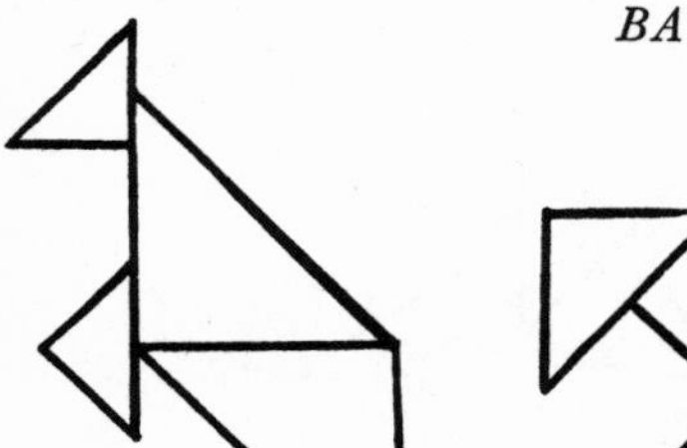

73 *Kangaroo*

74 *Horse*

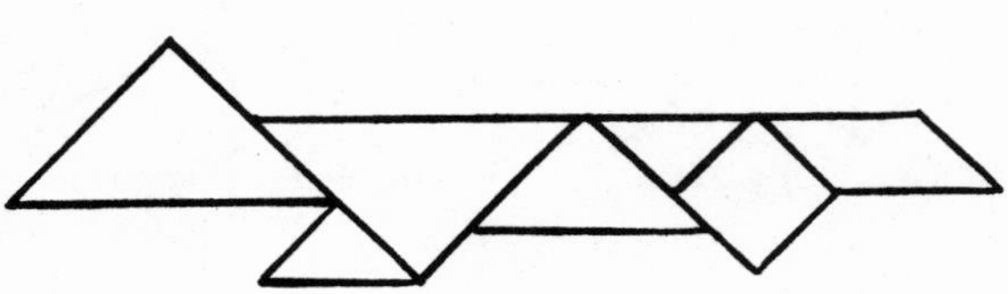

75 *Crocodile*

SEA CREATURES

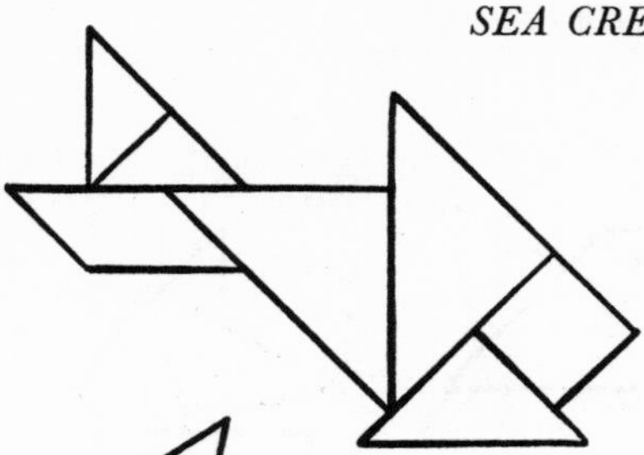

76

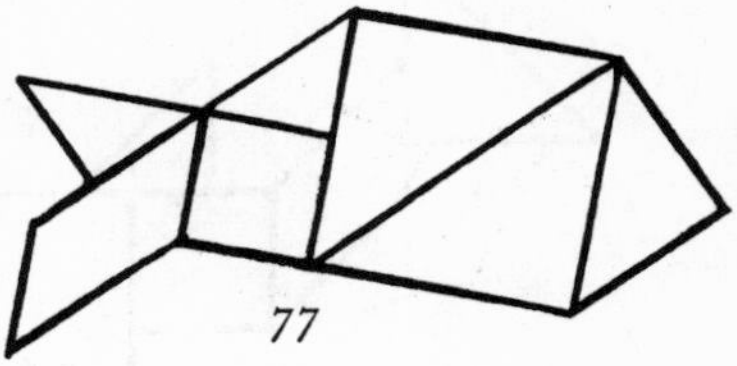

77

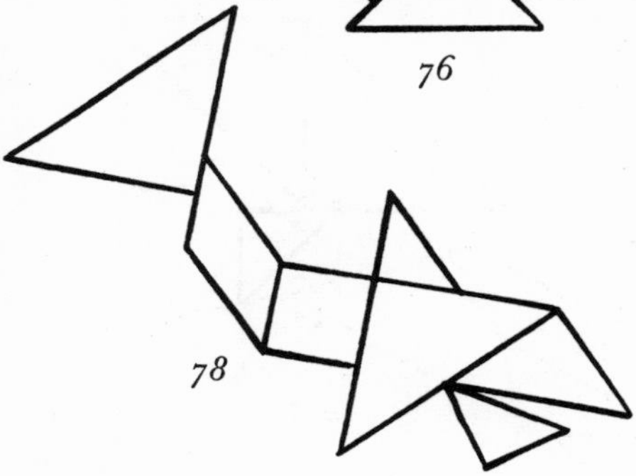

78

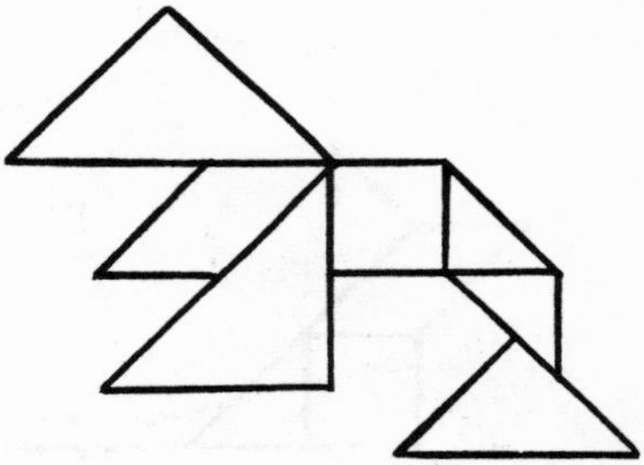

79 *Lobster*

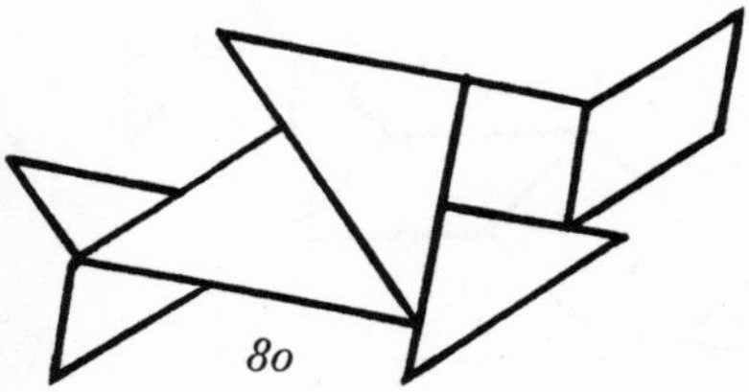

80

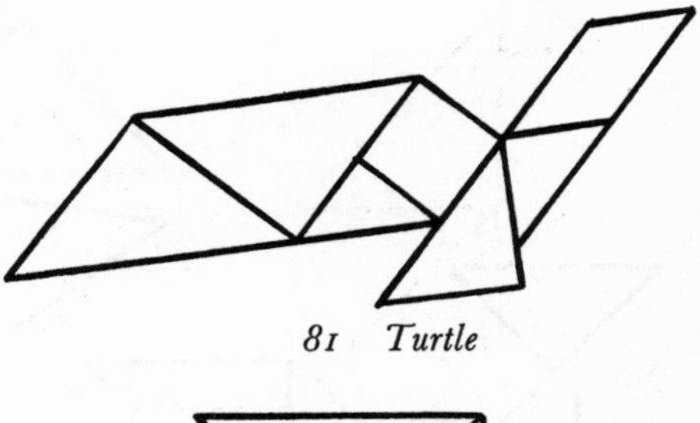

81 *Turtle*

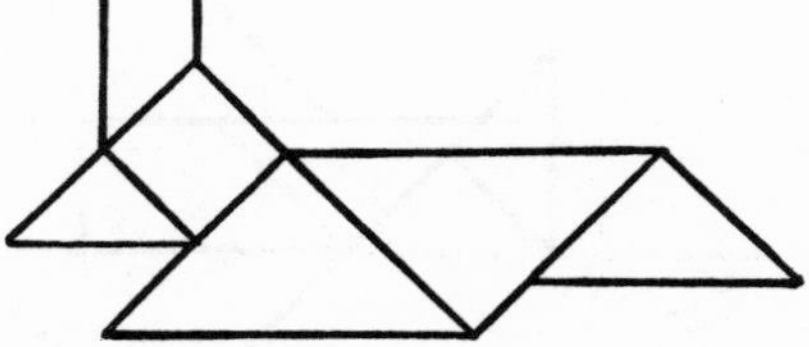

82 *Seal*

83 *Shrimp*

SEA CREATURES

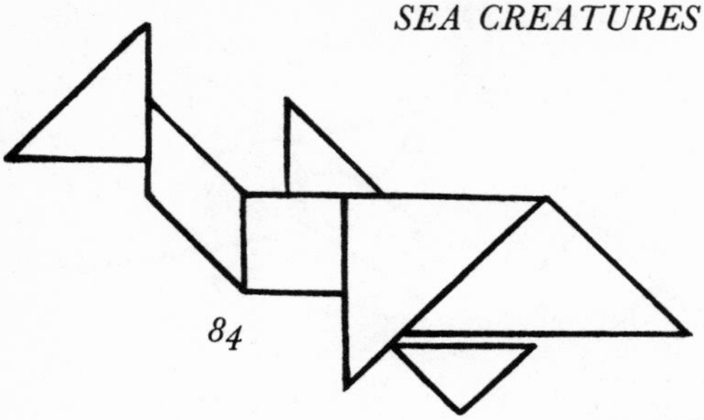

84

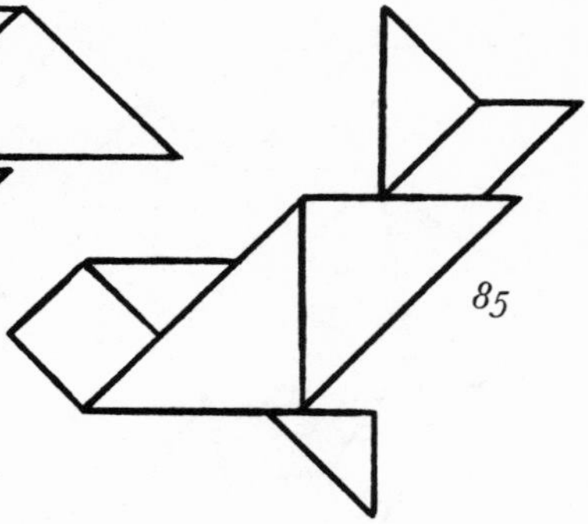

85

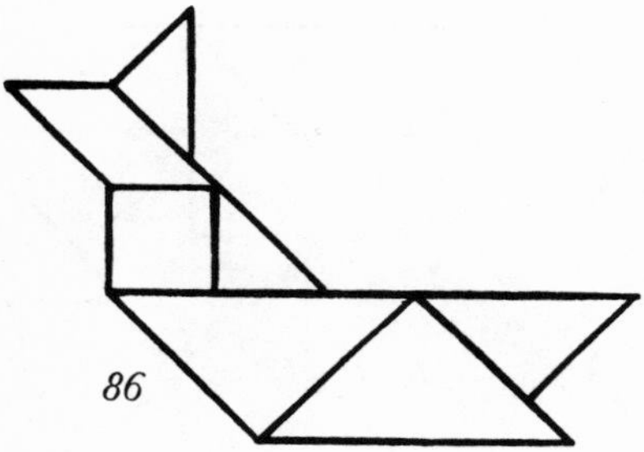

86

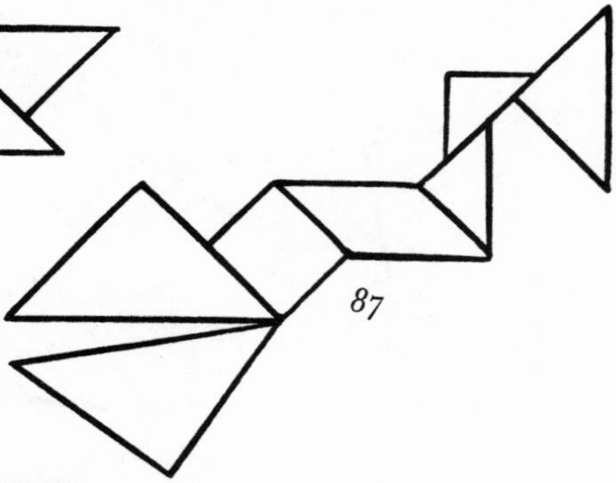

87

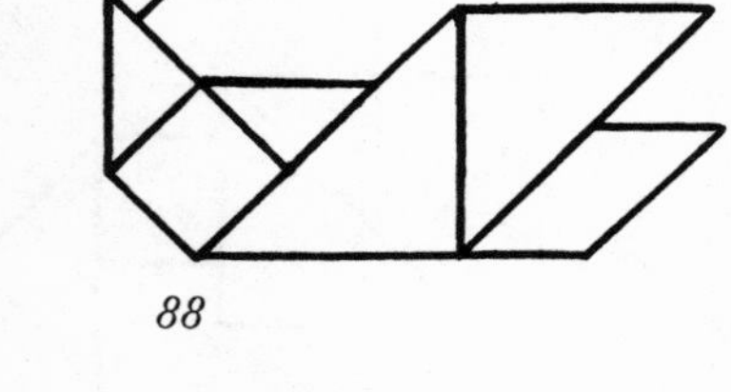

88

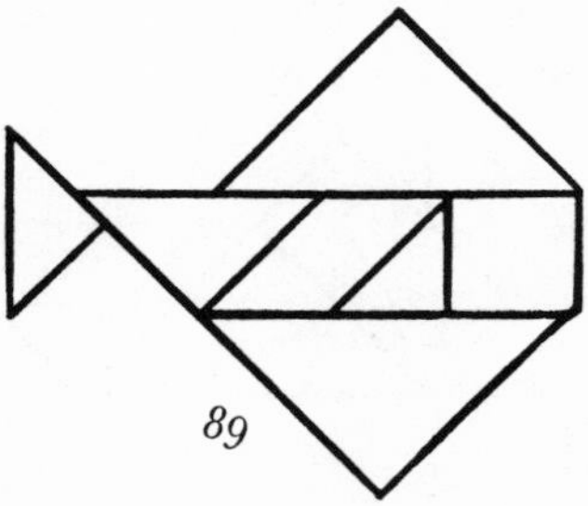

89

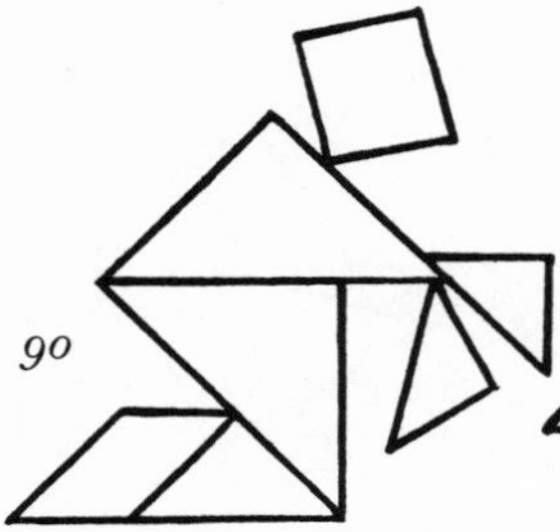

90

91

Shoeshine Boy and Customer

92

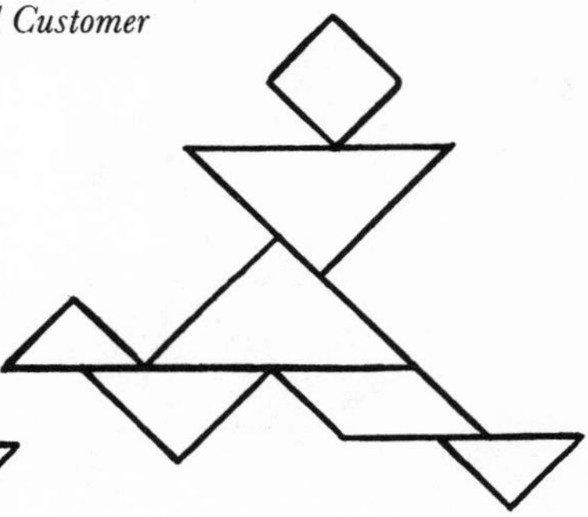

93

94

ACROBATS

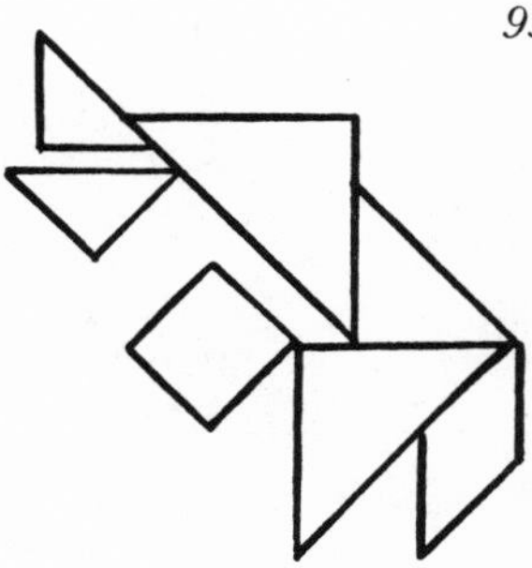

95

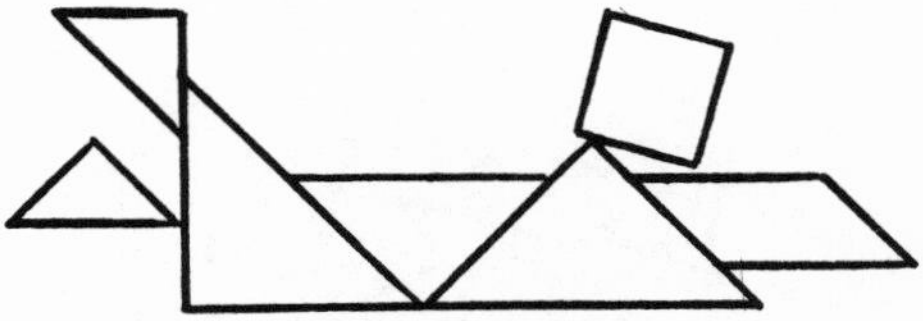

96

97

98

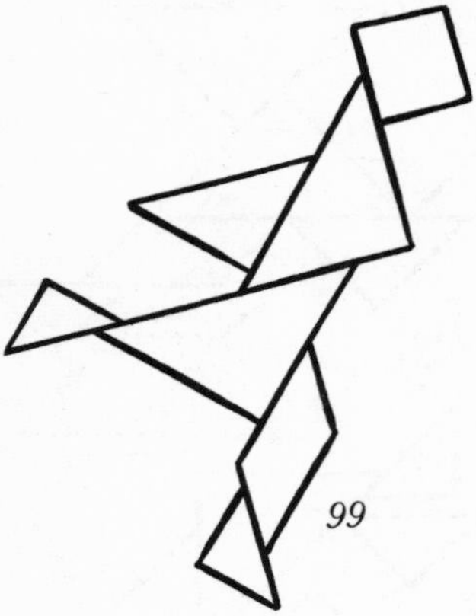

99

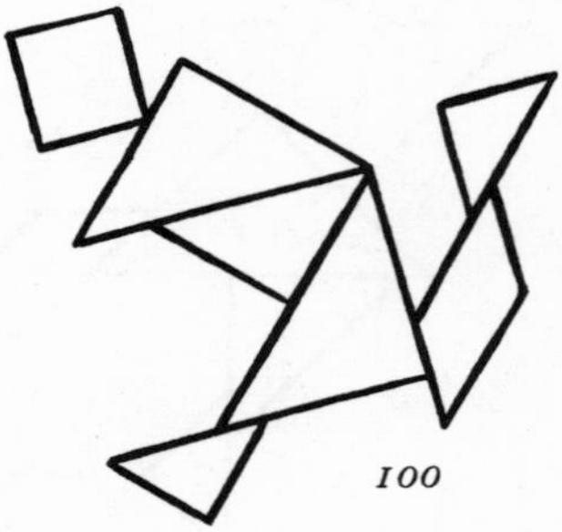

100

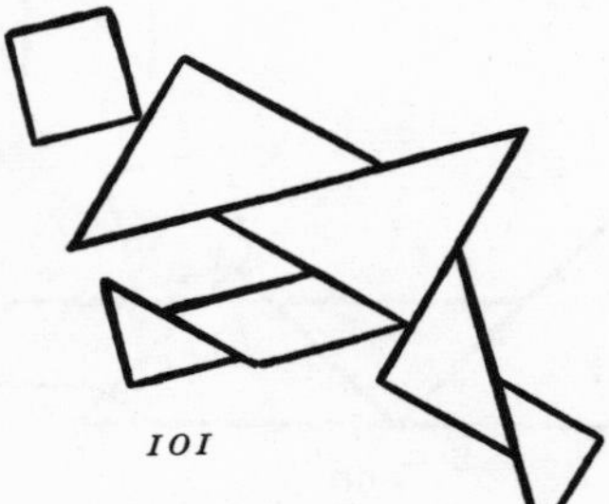

101

102

MEDIEVAL LADIES

103

104

105

106

AND SERVANTS

107

108

SAM LOYD'S PORTRAIT GALLERY

109

110

111

112

113

114

115

116

117

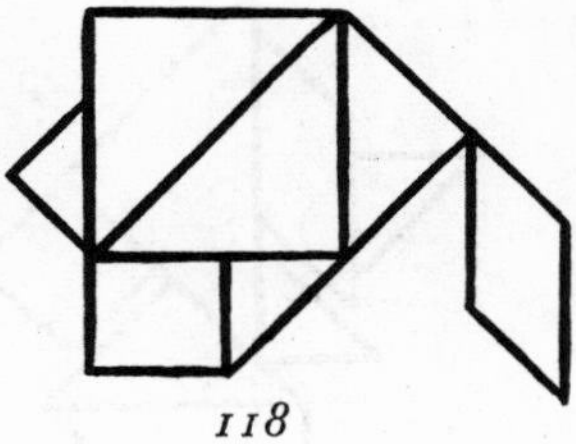

118

119

120

121

122

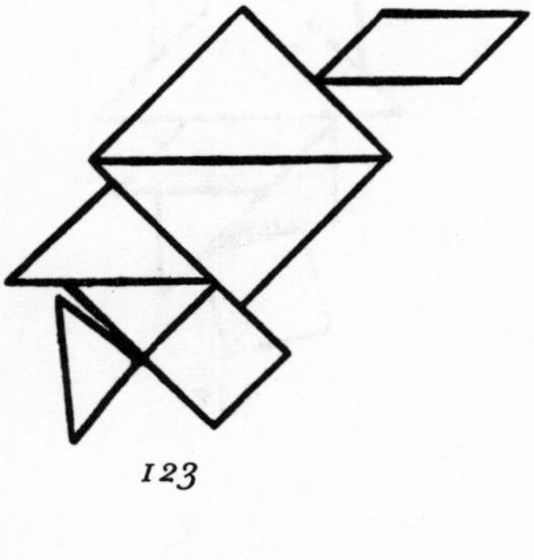

123

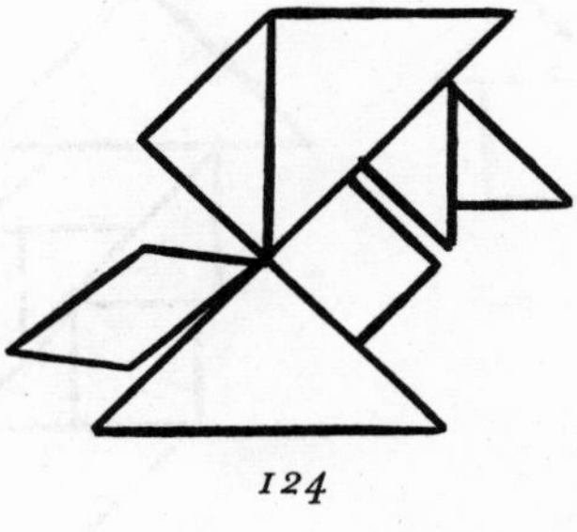

124

125

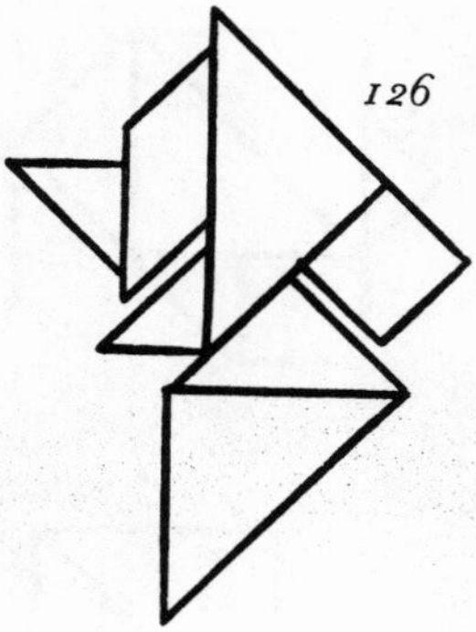

126

127

128

129

130

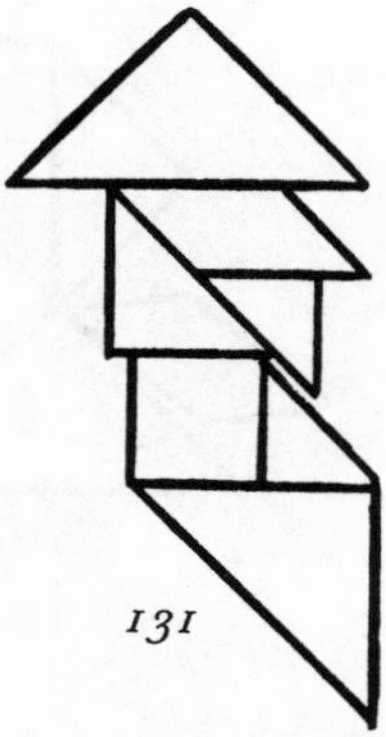

131

132

133

134

135

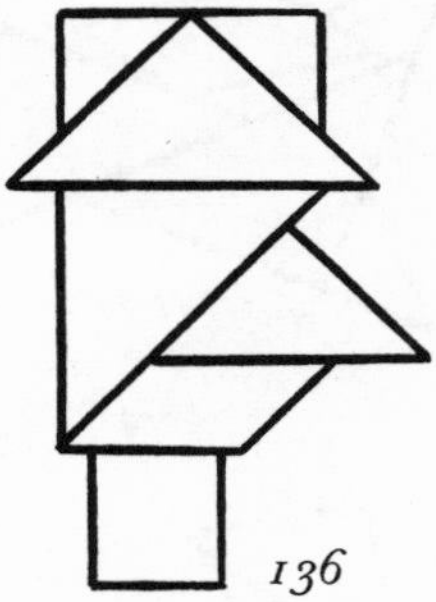

136

137

138

HORSEMEN

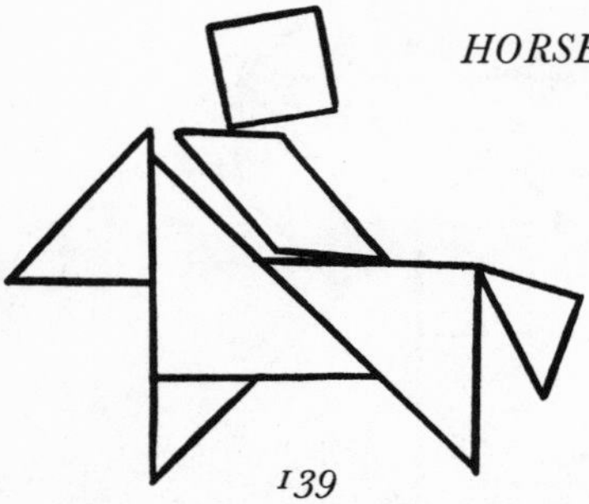

139

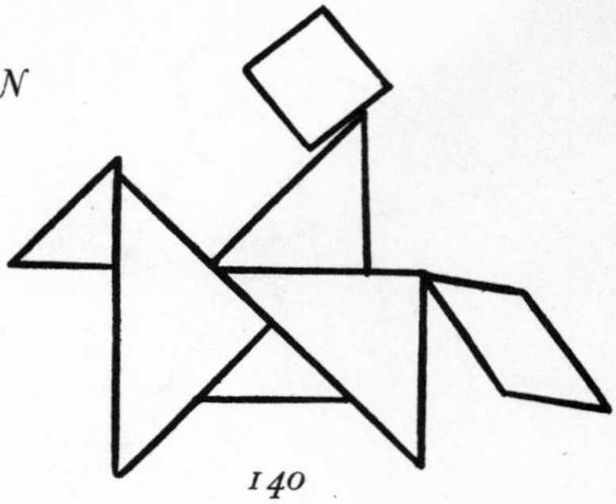

140

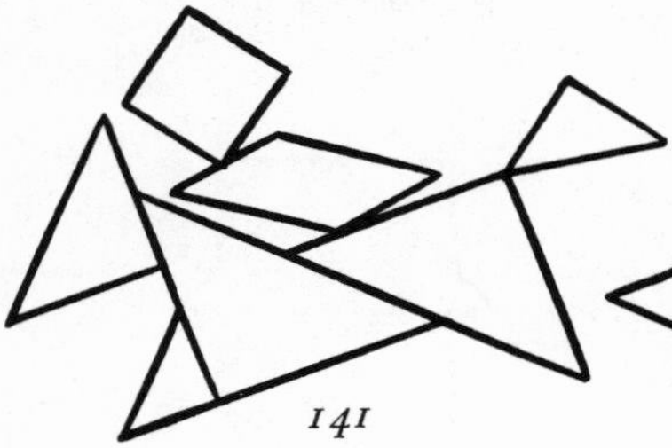

141

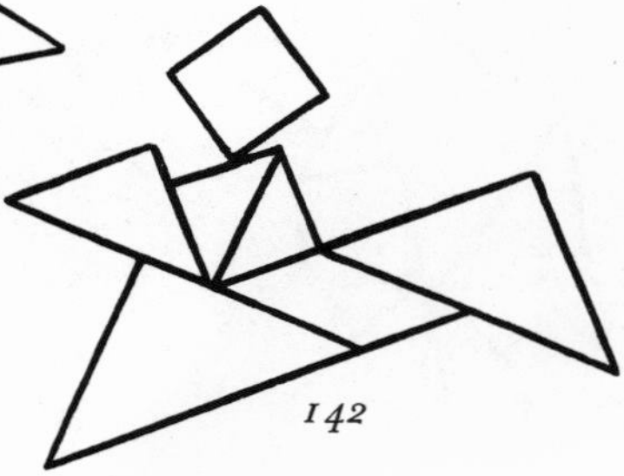

142

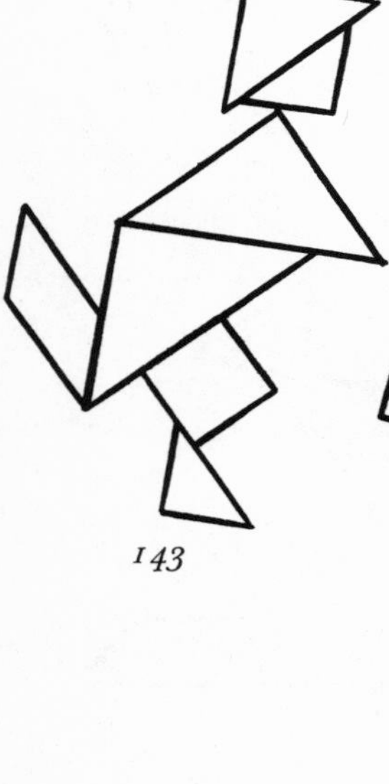

143

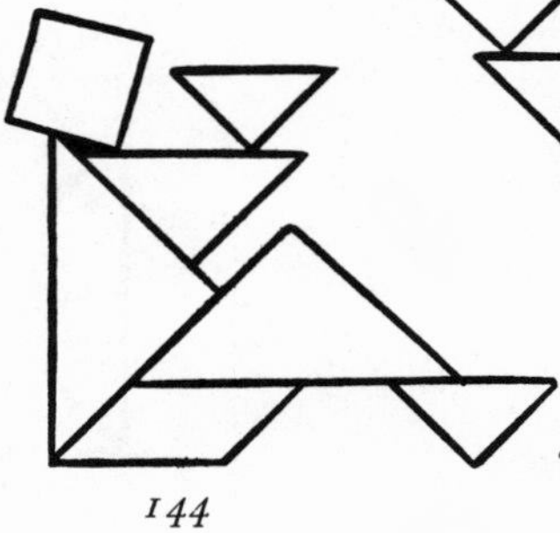

144

145

146 *Pipe*

147

148 *Candle*

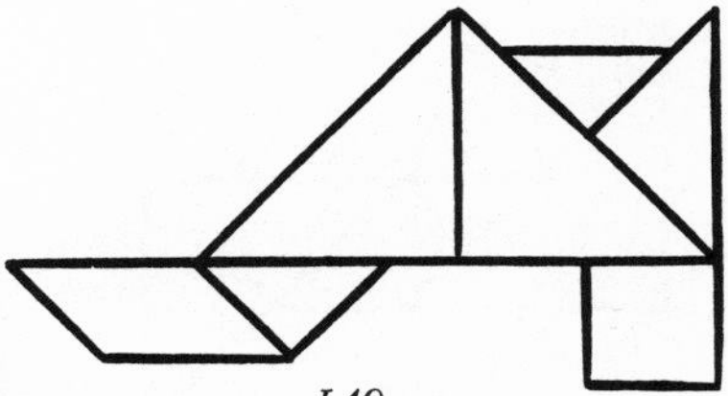

149

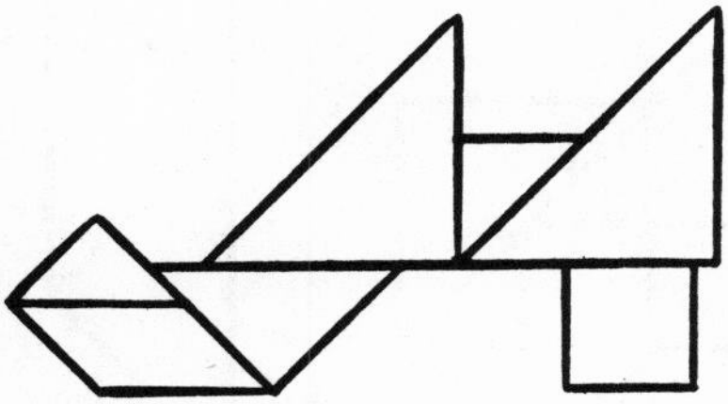

150

151

152

SHOES

153

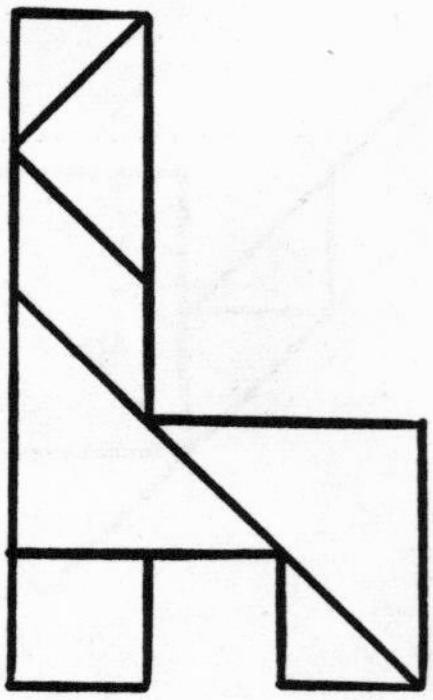

154

CHAIRS

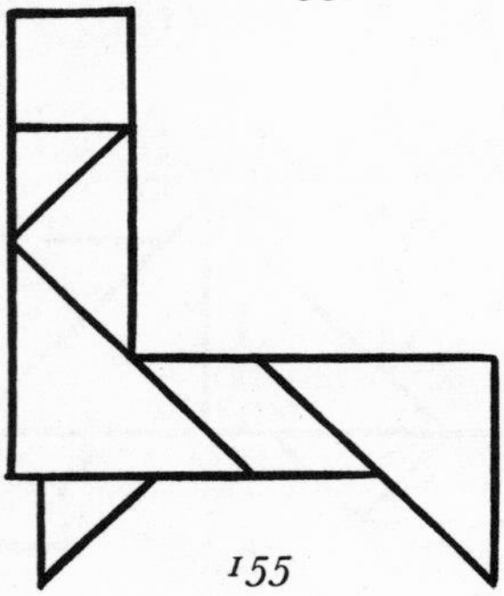

155

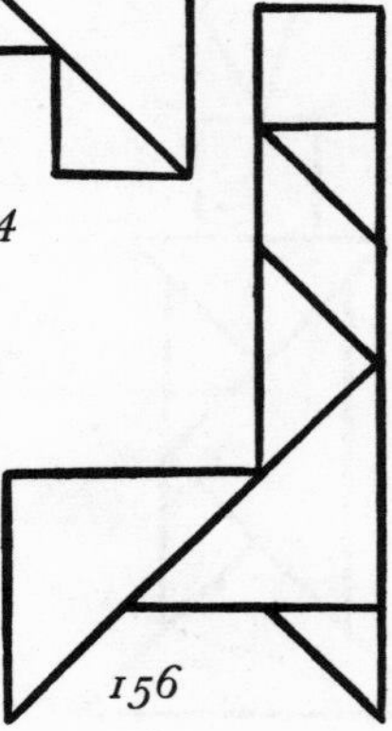

156

157

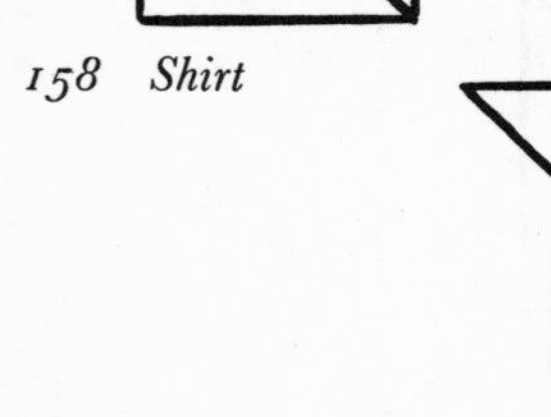

158 Shirt

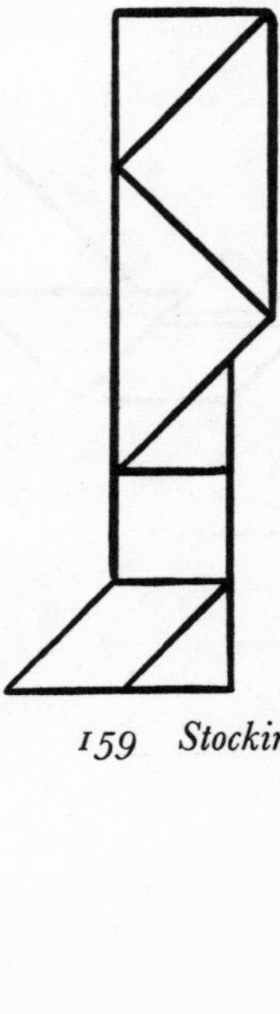

159 Stocking

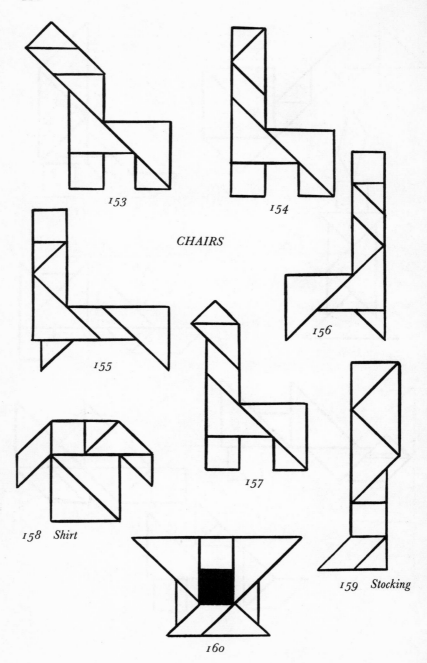

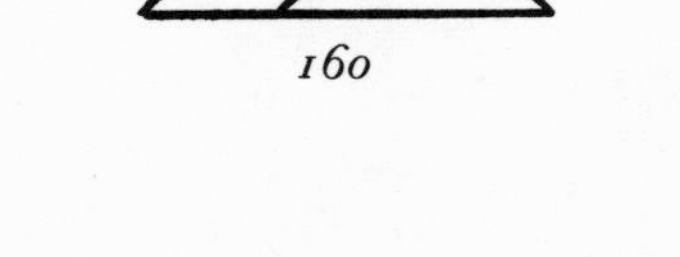

160

161 *Watering can*

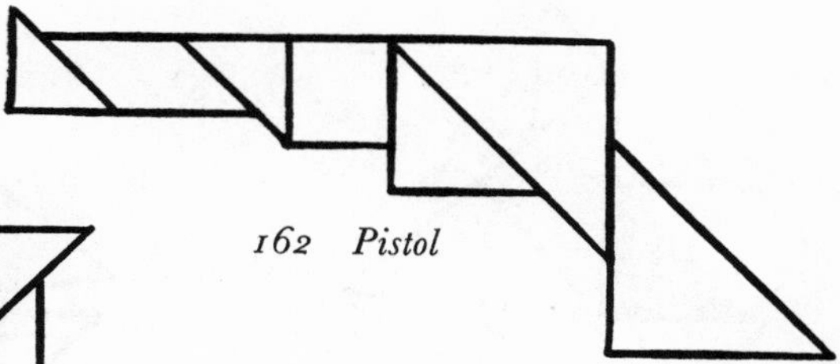

162 *Pistol*

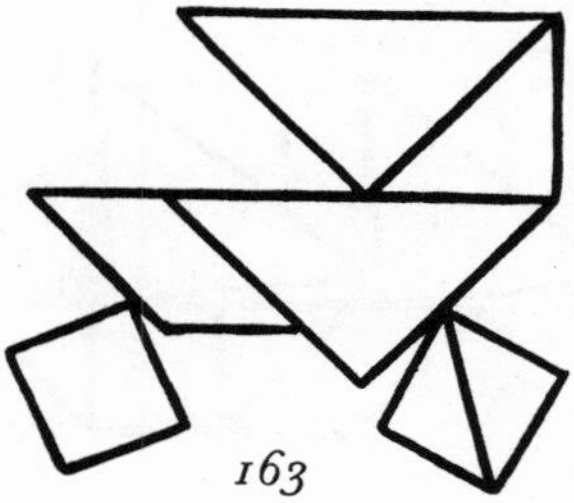

163

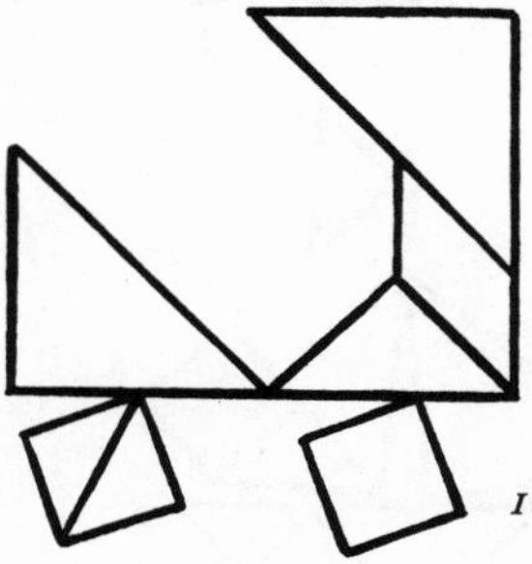

164

BABY CARRIAGES

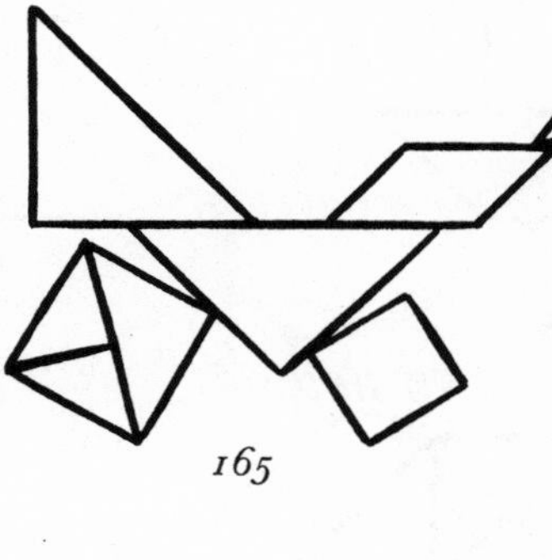

165

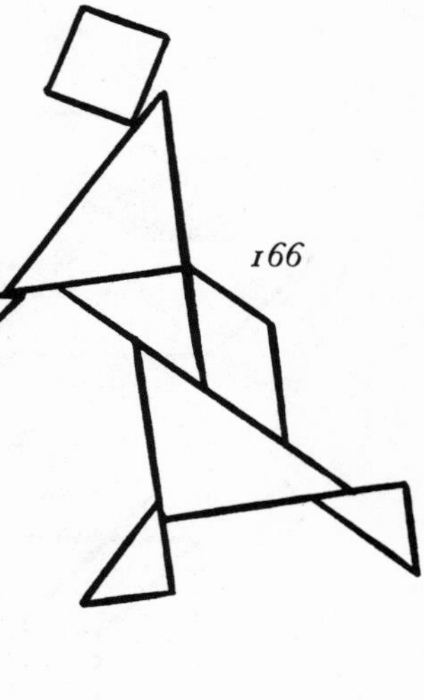

166

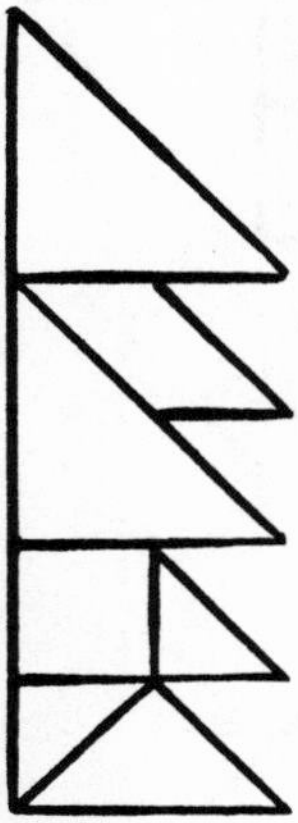

167 ???

BOATS

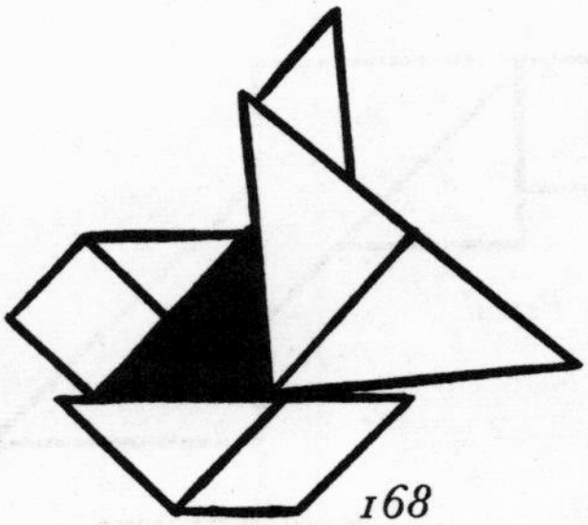

168

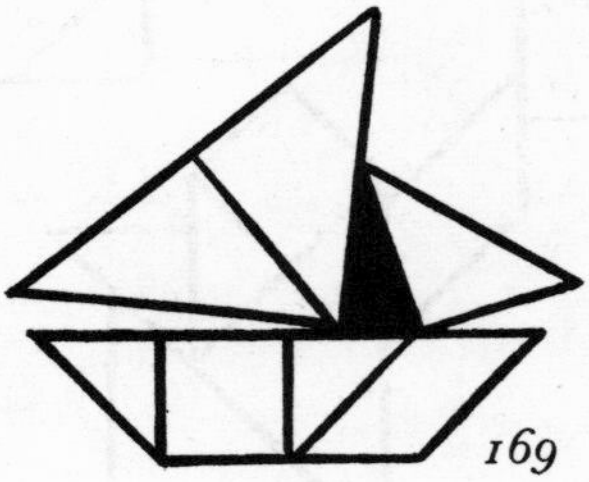

169

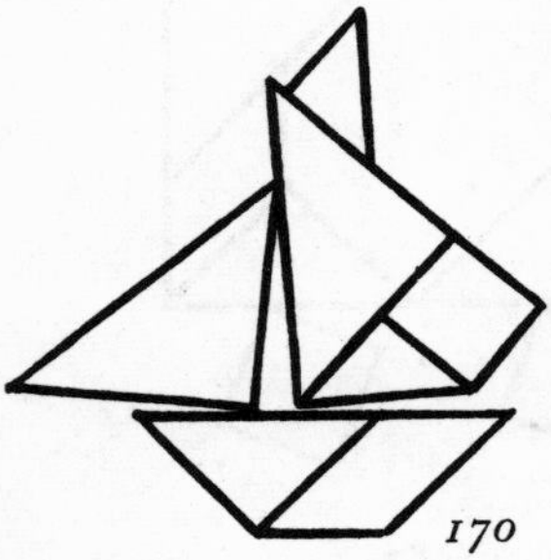

170

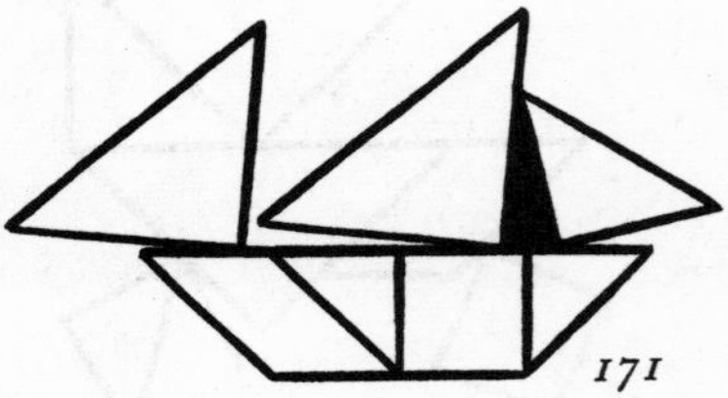

171

172

173

AND BRIDGES

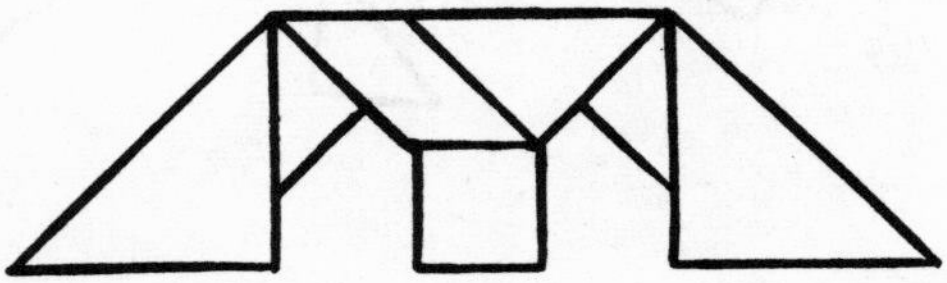

174

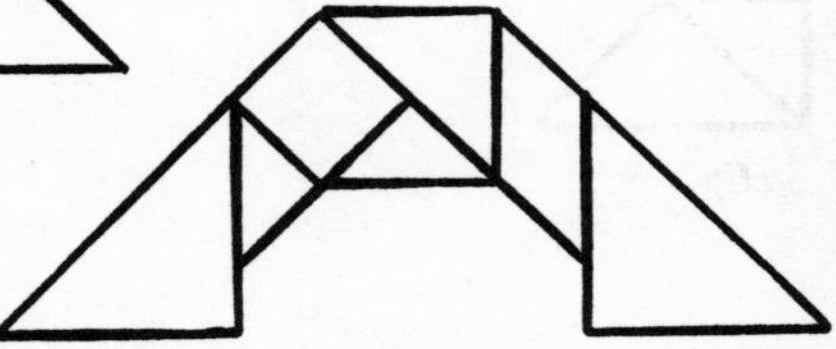

175

BRIDGES

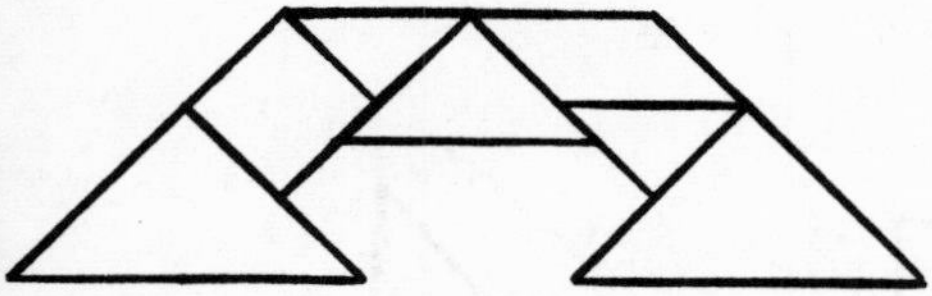

176

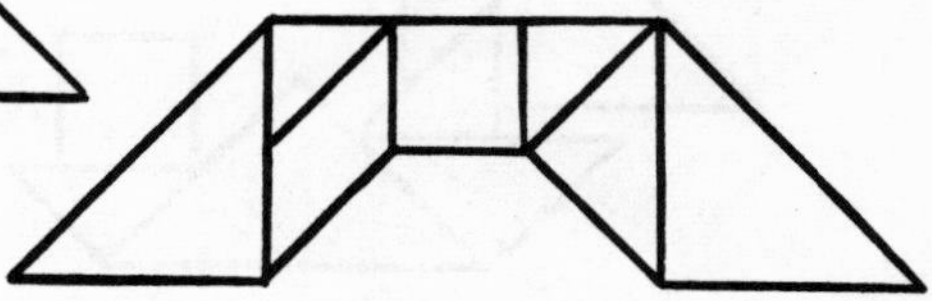

177

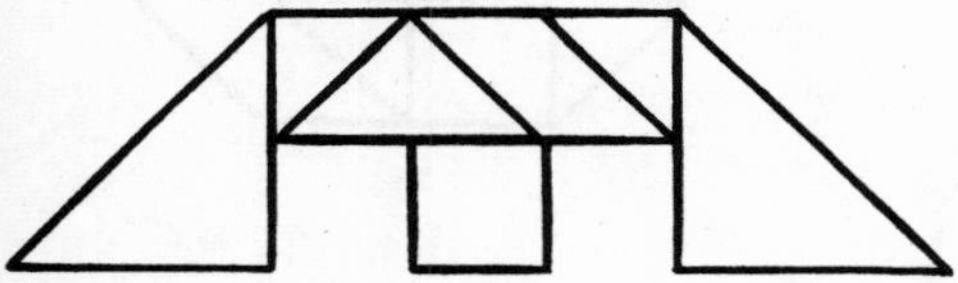

178

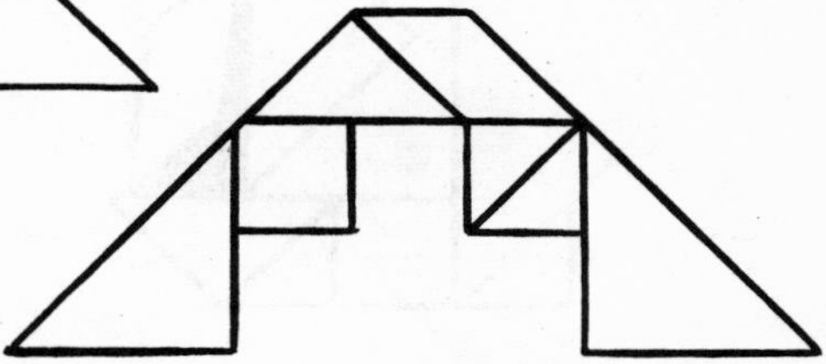

179

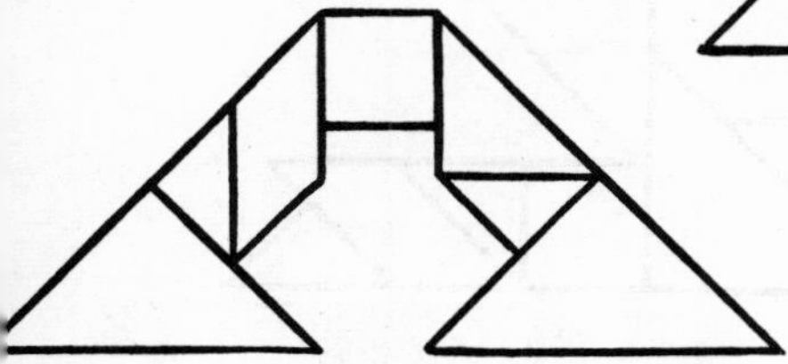

180

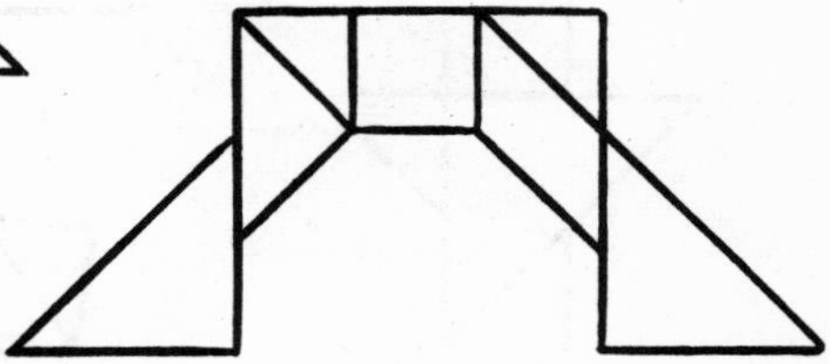

181

182

BOATS

183

184

185

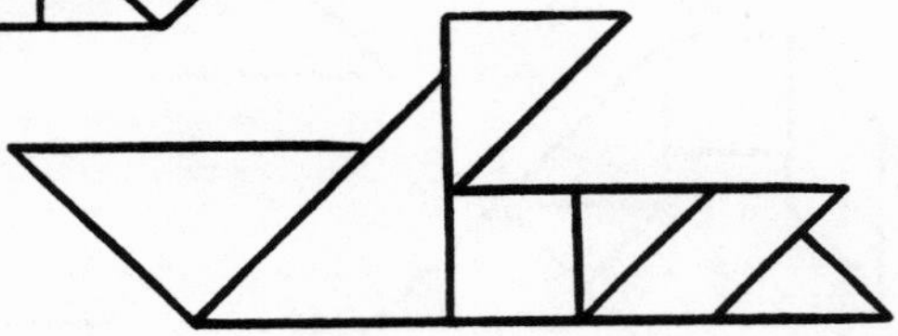

186

187 & 188 Rowing to the lighthouse

THE HOUSE THAT JACK BUILT

Illustrated by Sam Loyd

189

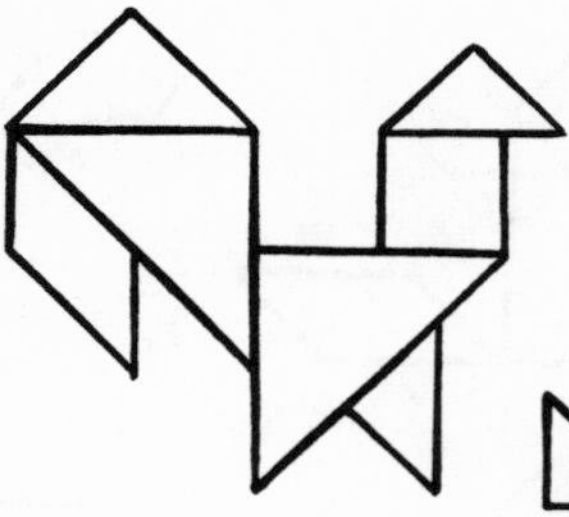

190

191

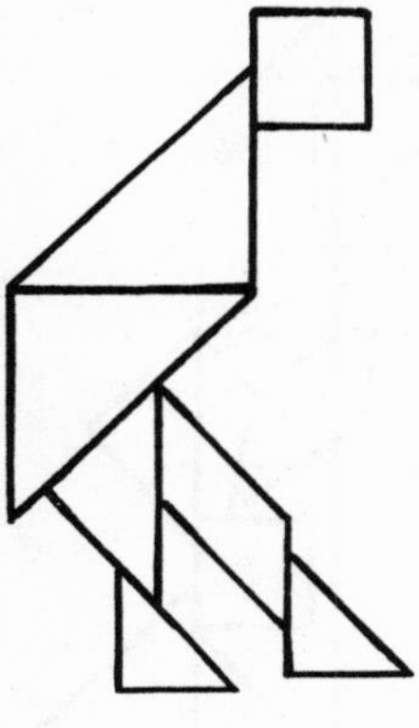

192

193

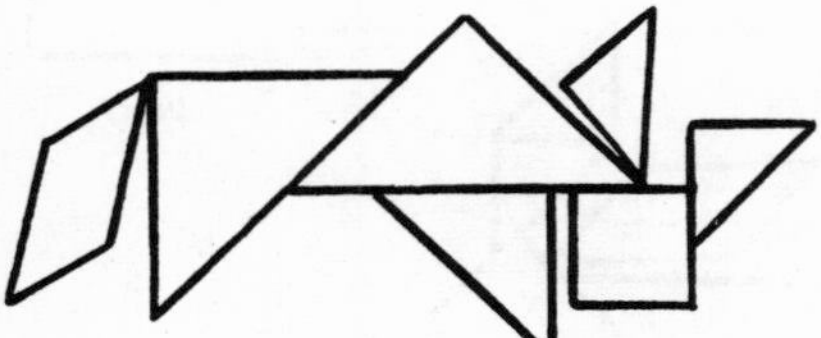

194

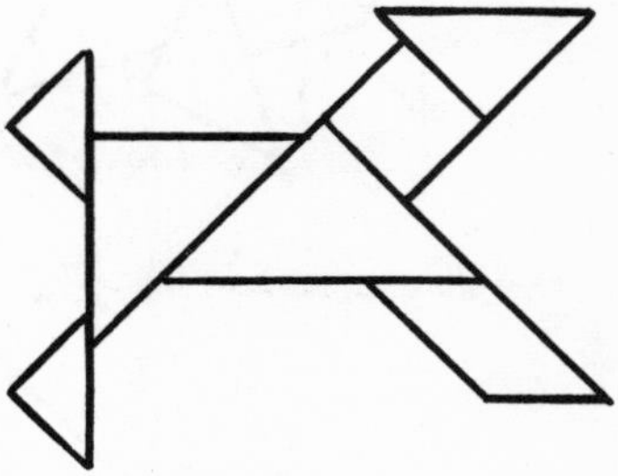

195

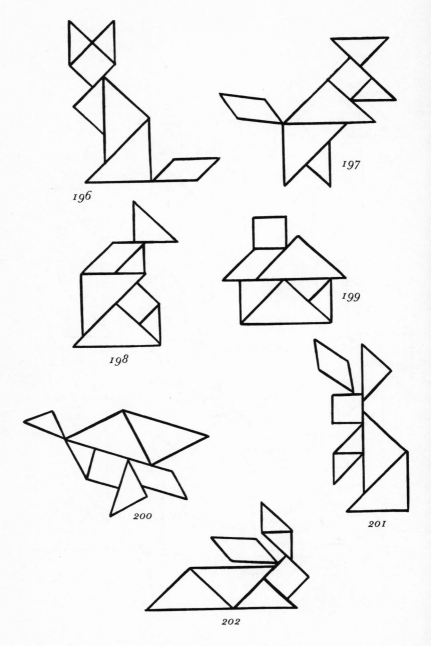

196

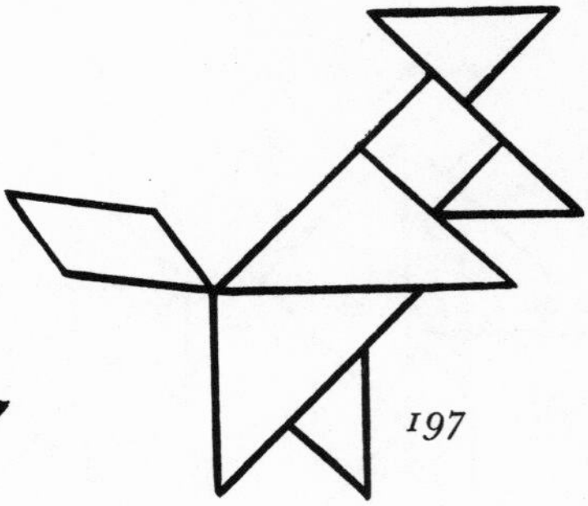

197

198

199

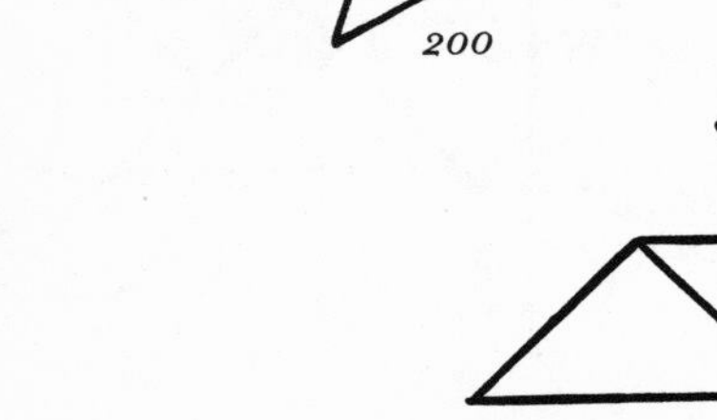

200

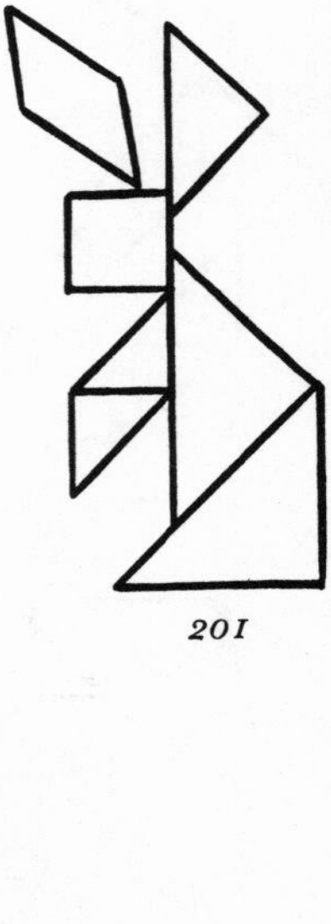

201

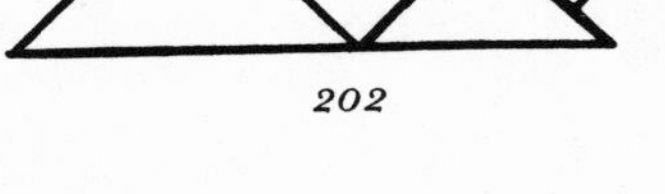

202

THE STORY OF CINDERELLA

Illustrated by Sam Loyd

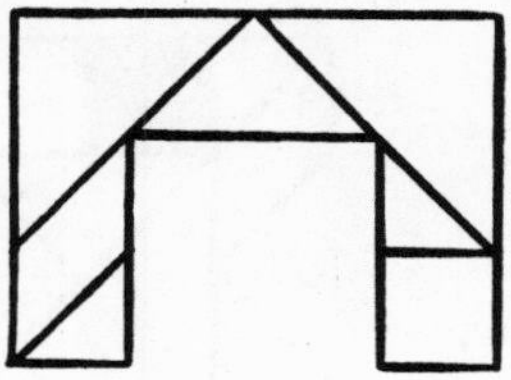

203

204

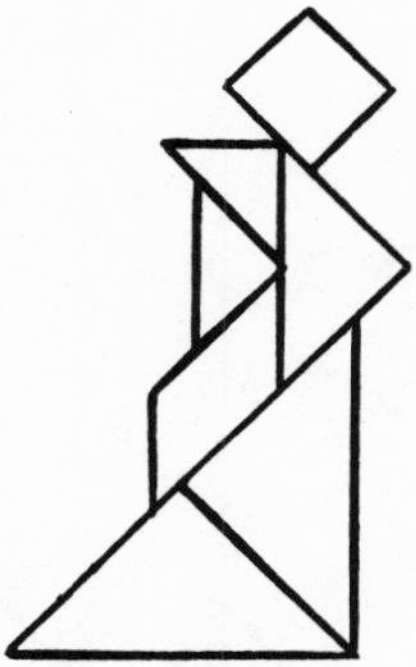

205

206

207

208

209

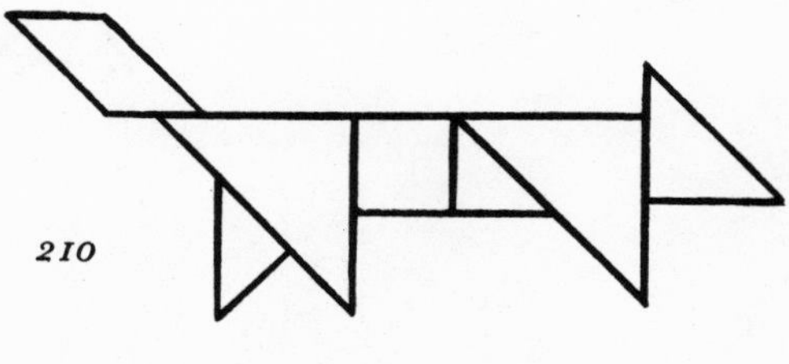

210

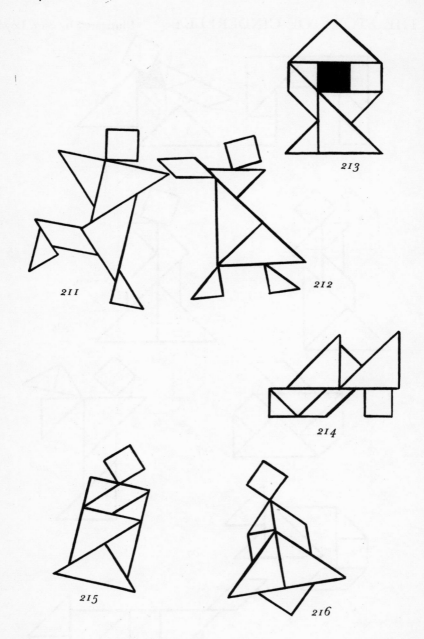

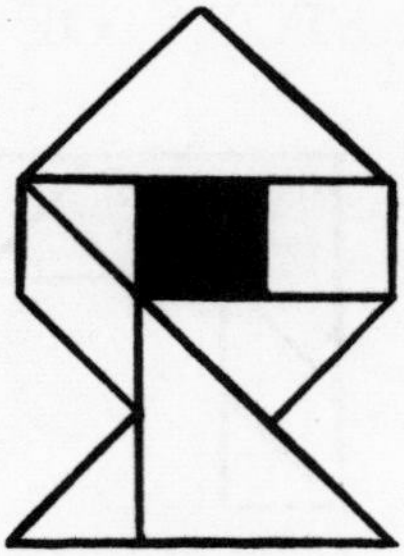

213

211

212

214

215

216

A GAME OF BILLIARDS

By H. E. Dudeney

217

218

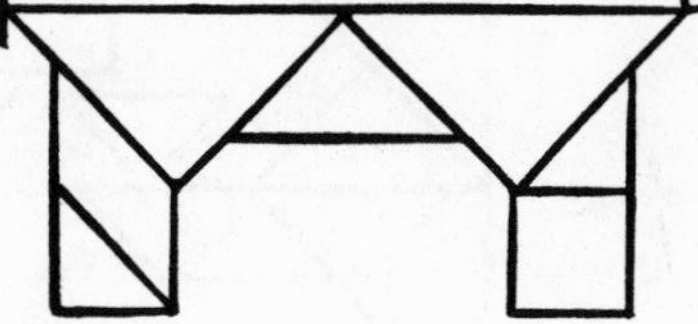

219

220

221

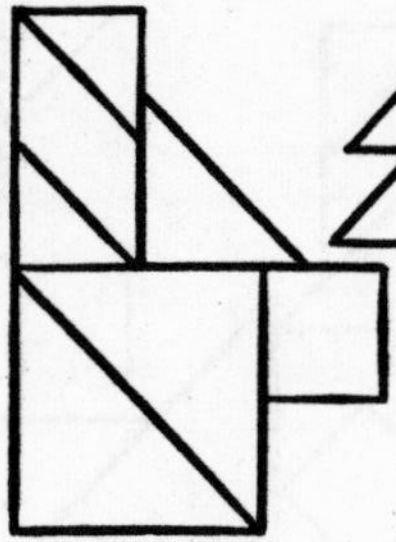

222

223

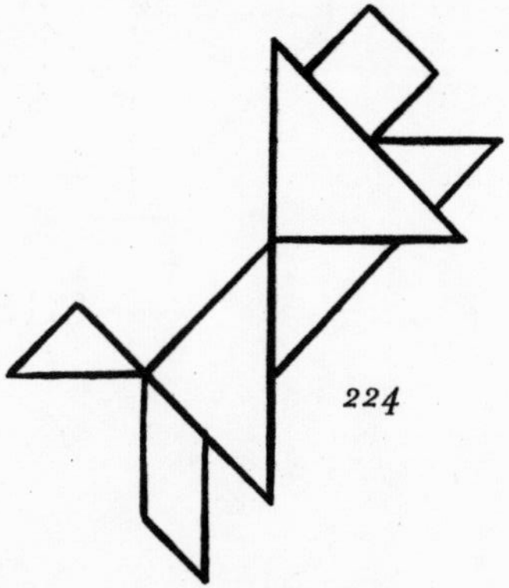

224

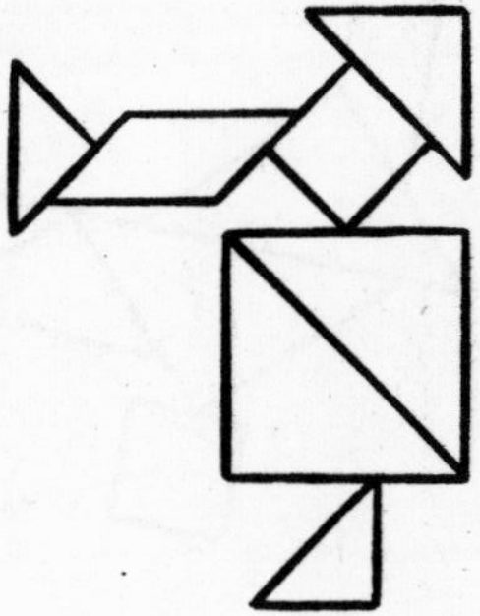

225

THE ORCHESTRA

By H. E. Dudeney

226

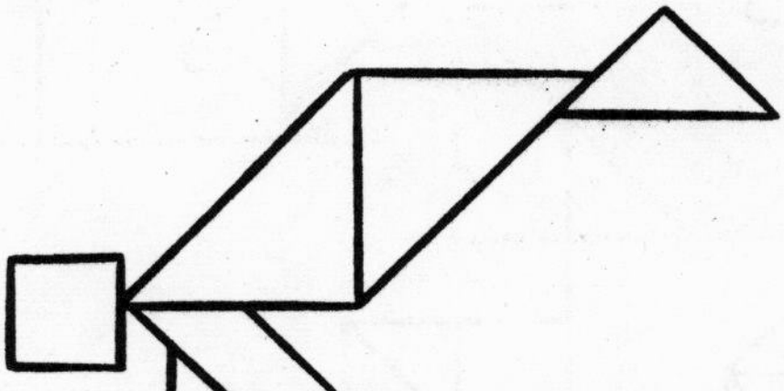

227

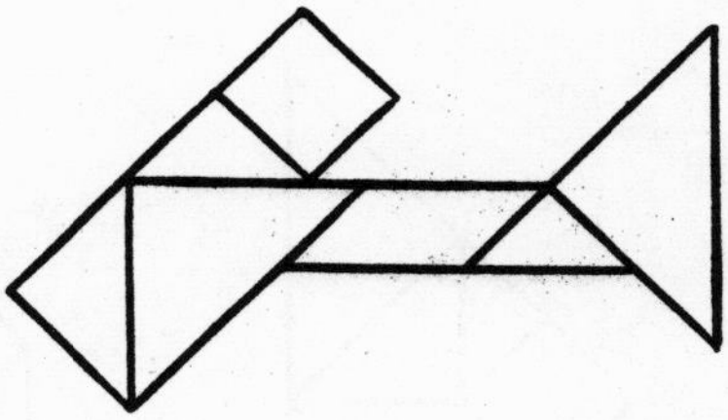

228

229

230

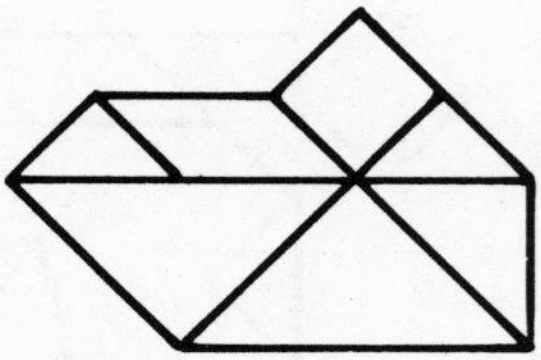

231

232

233

234

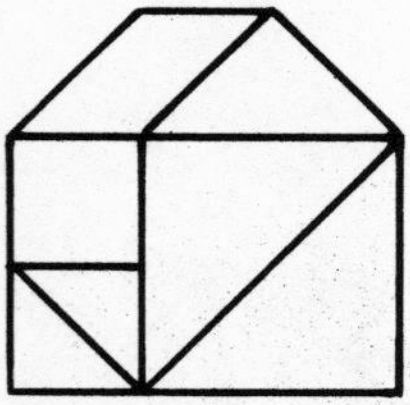

235

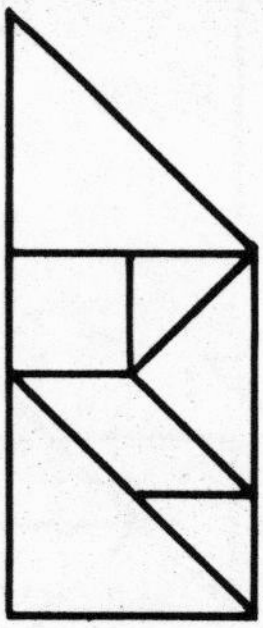

236

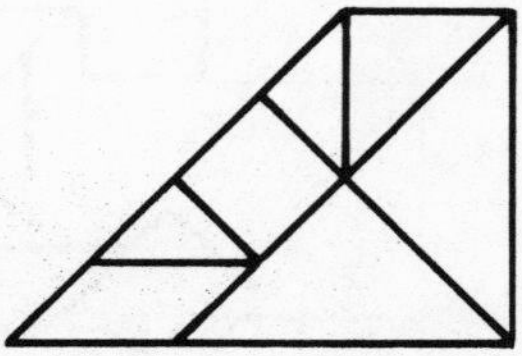

237

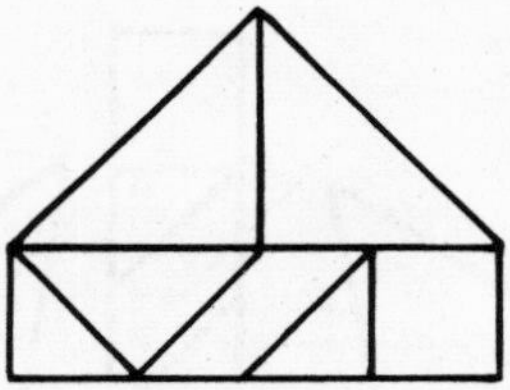

238

239

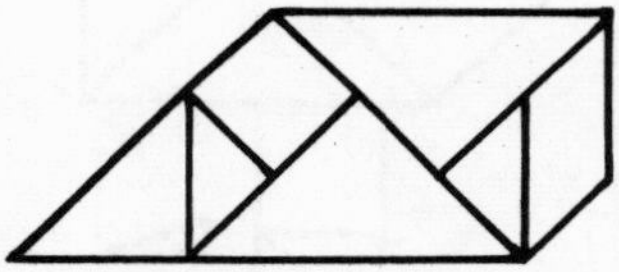

240

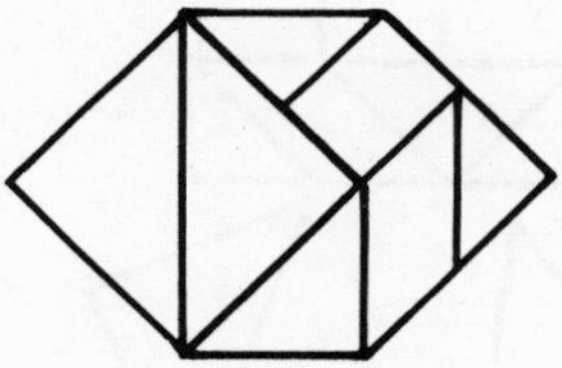

241

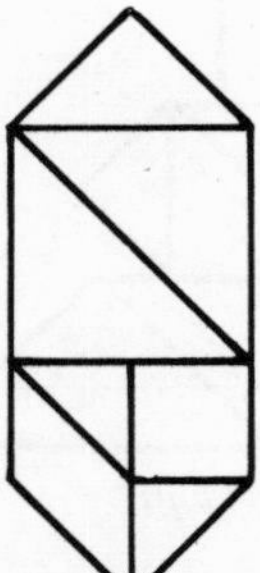

242

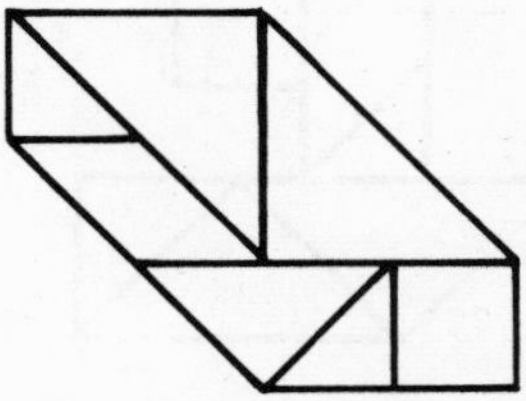

243

SOME CHINESE CHARACTERS

244

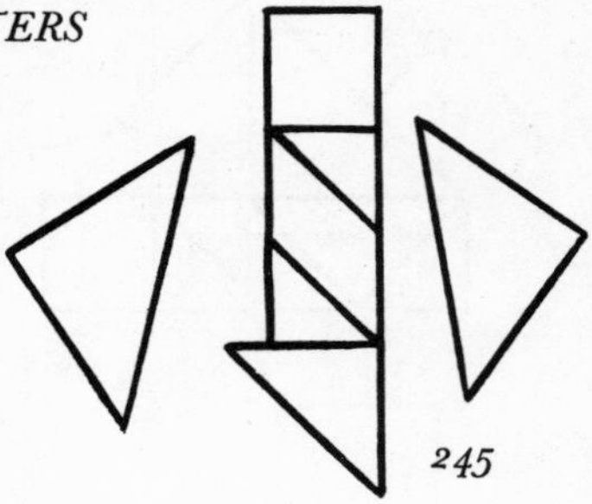

245

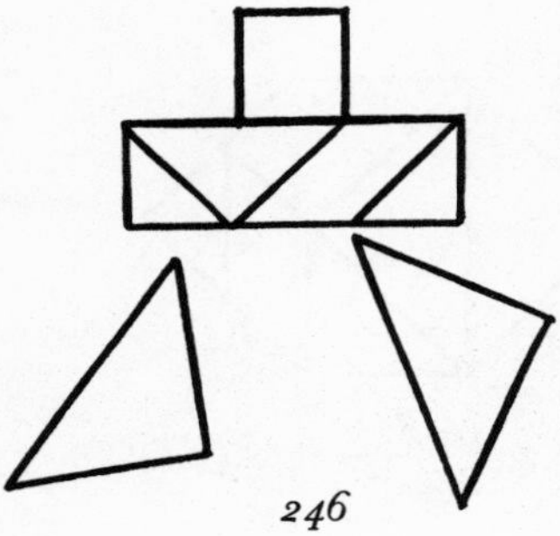

246

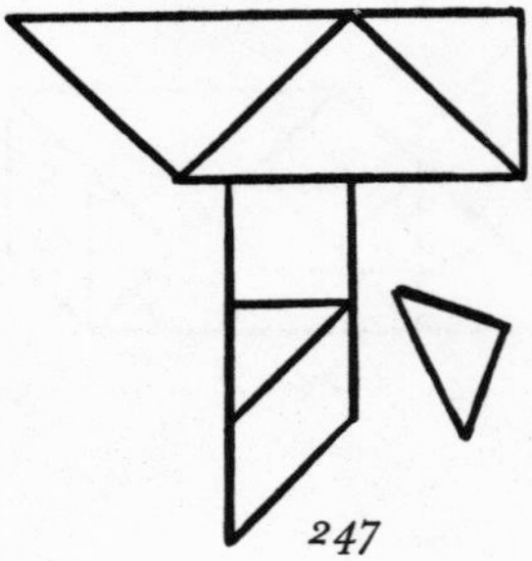

247

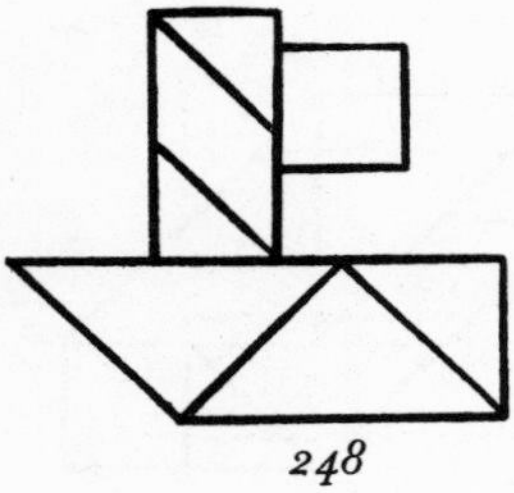

248

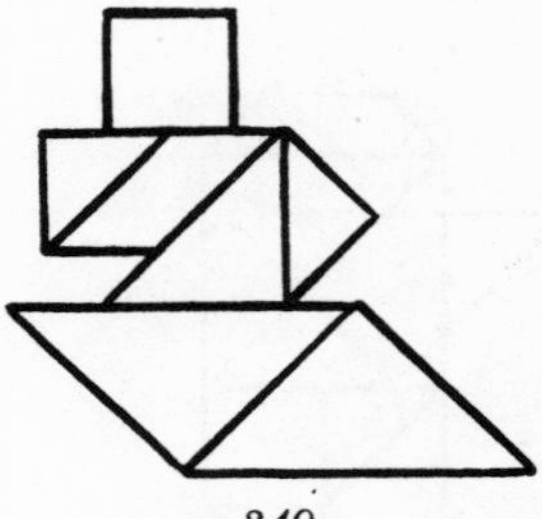

249

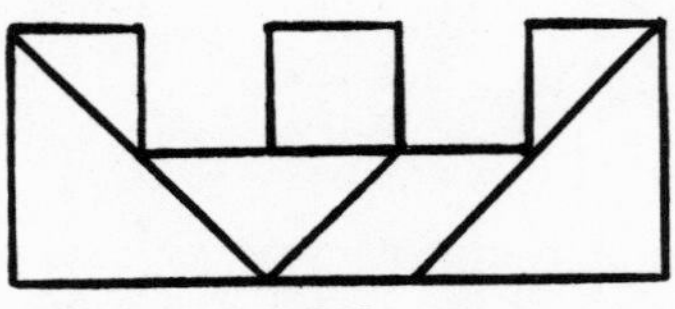

250

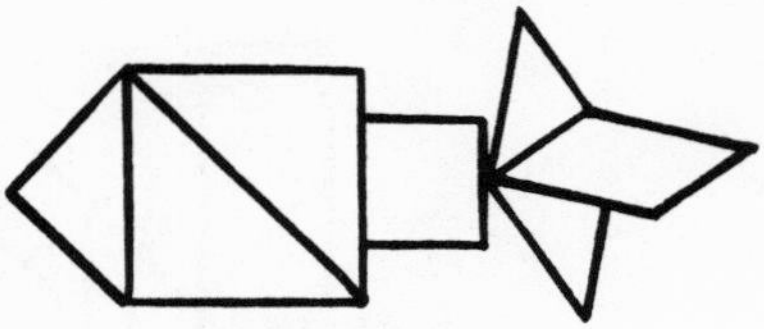

251 Goldfish

252 Riding a crane

253 Bottle gourd

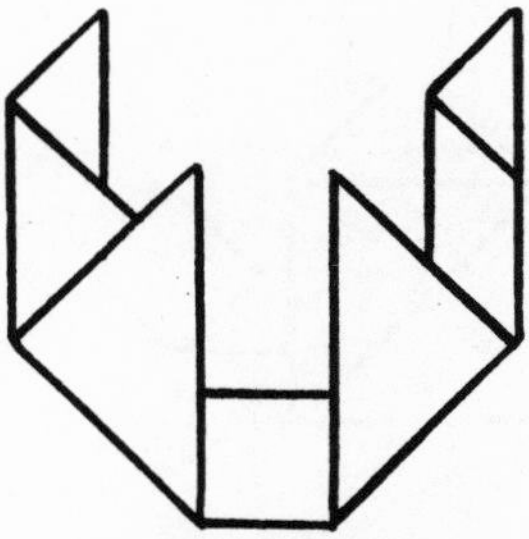

254 "Buddha's hand"

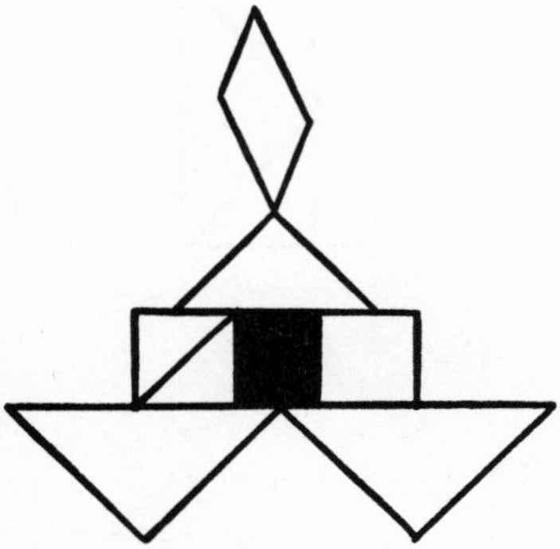

255 Helmet

256 "Drinking alone"

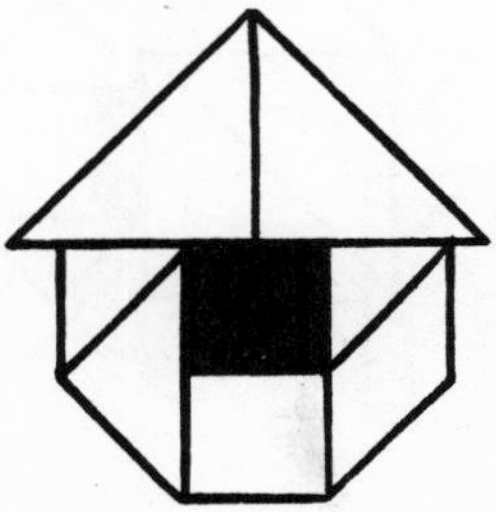

257 Clock

258 Brazier

259 Dagger

260 Pinking knife

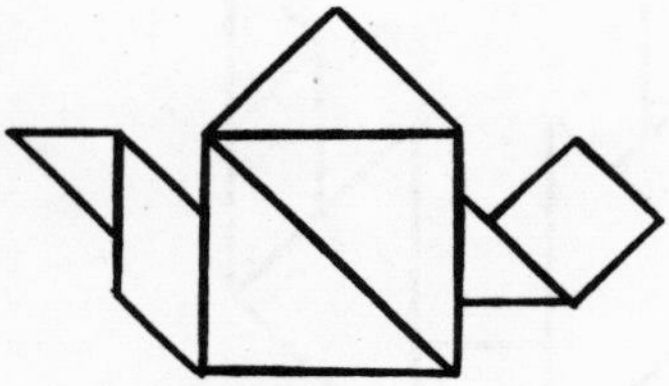

261 Teapot

262 Mulberry hook

263 Raincoat

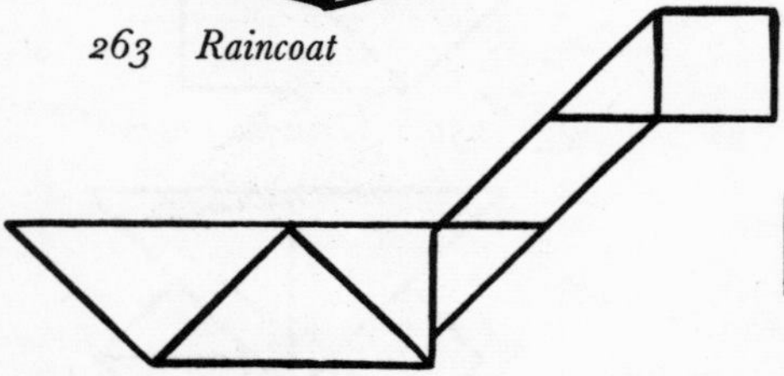

264 Soup spoon

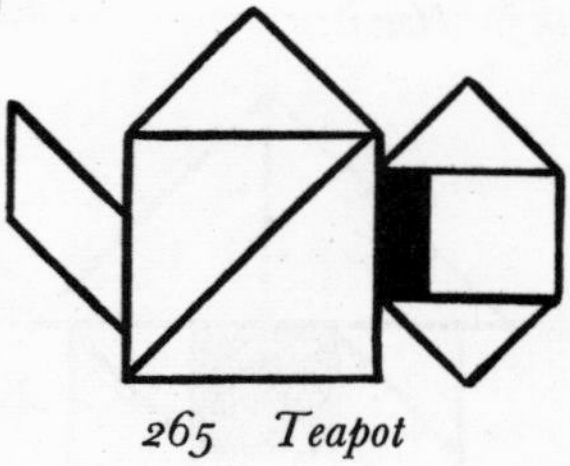

265 Teapot

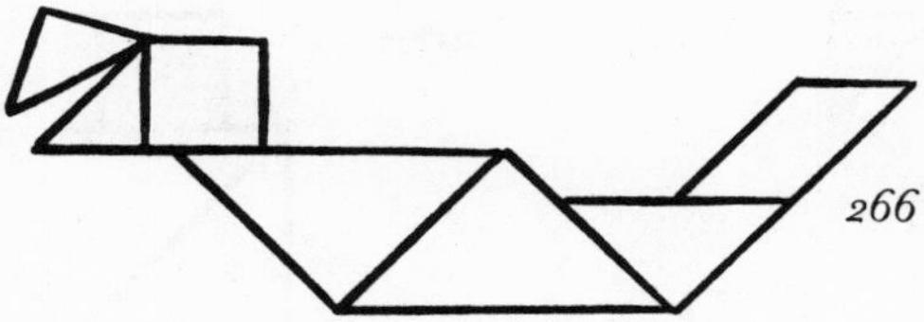

266 *Dragon boat*

267 *Three-legged toad*

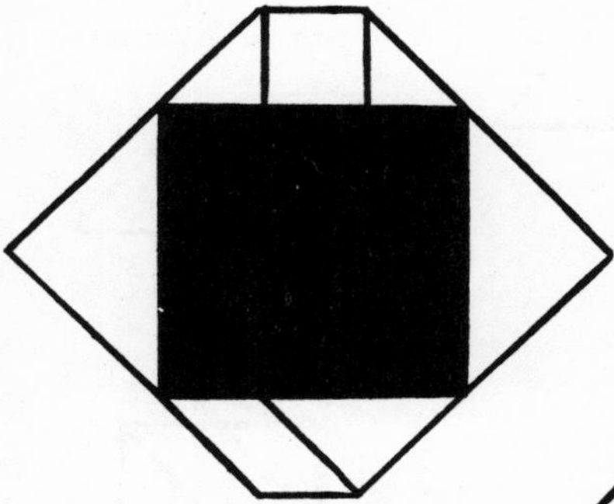

268 *Pendant*

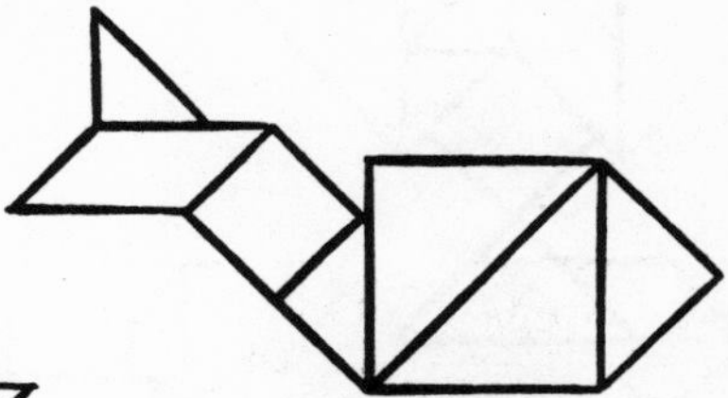

269 *Carp*

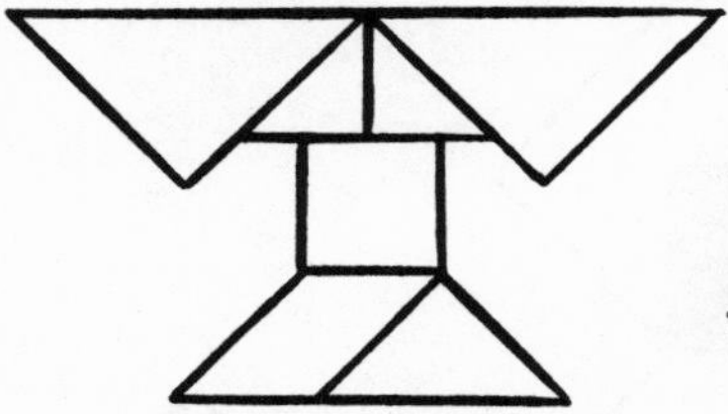

270 *Flower bowl*

271 *Scissors*

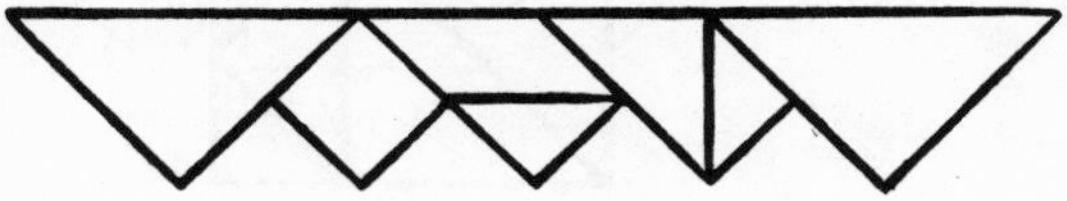

272 *Saw*

273 Clothes post

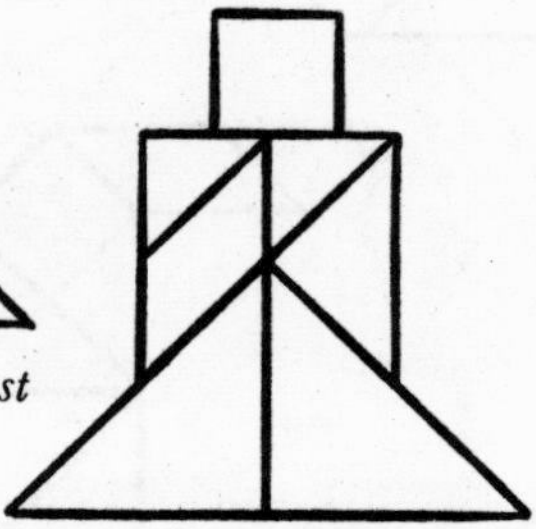

274 Bell

275 Feather tube

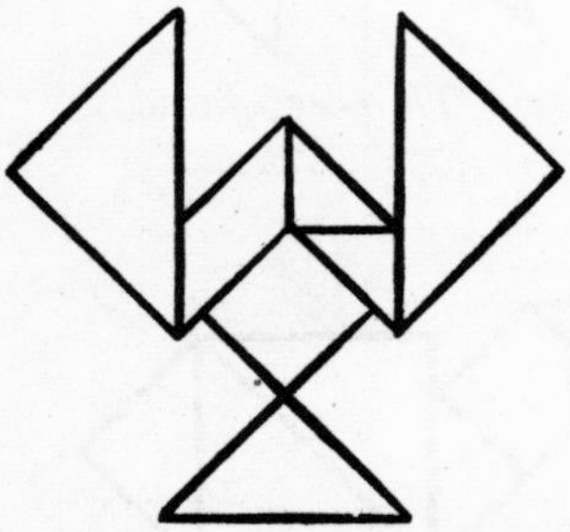

276 Lucky charm

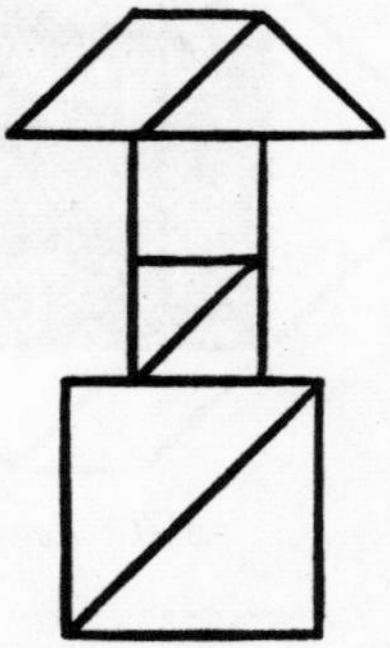

277 Look-out tower

278 Medicine jar

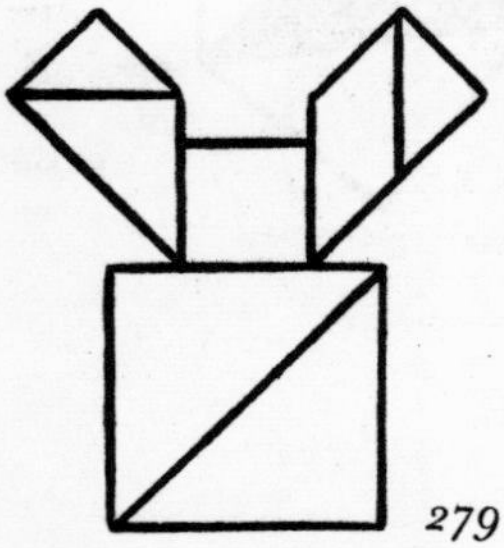

279 Mirror rack

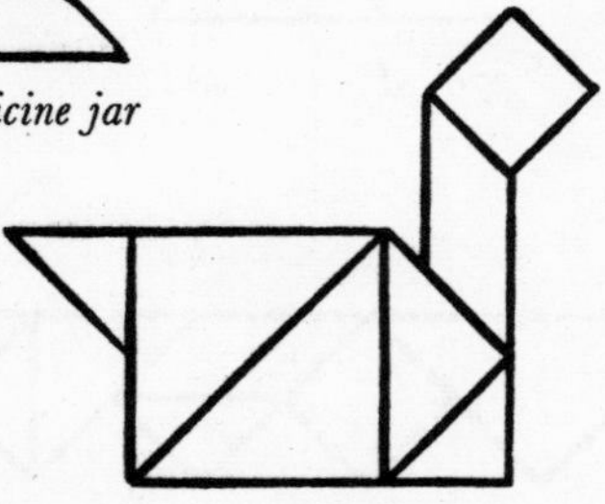

280 Ladle

TWO BASKETS

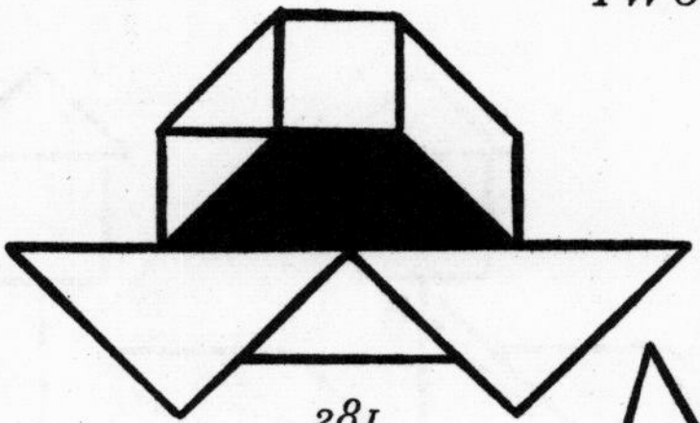

281

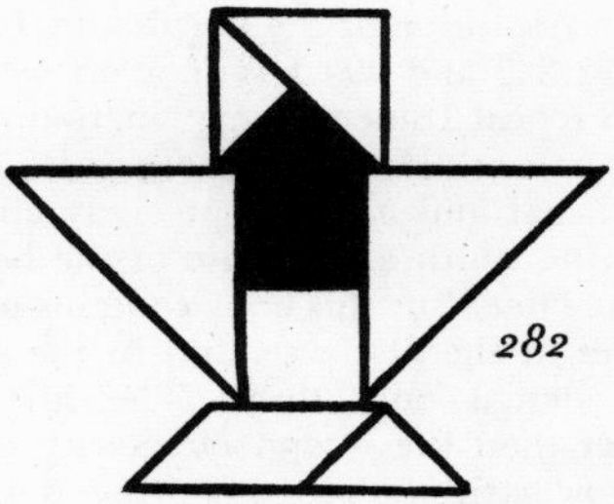

282

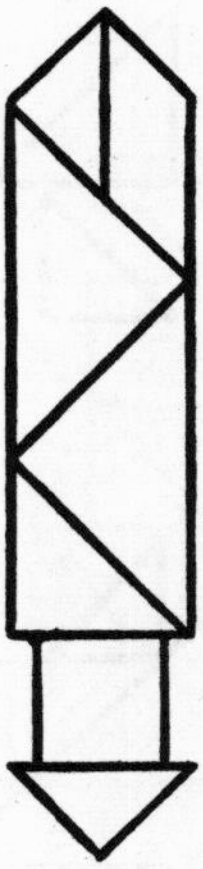

283 Arrow

284 Candle

285 Water pot

286 Measure

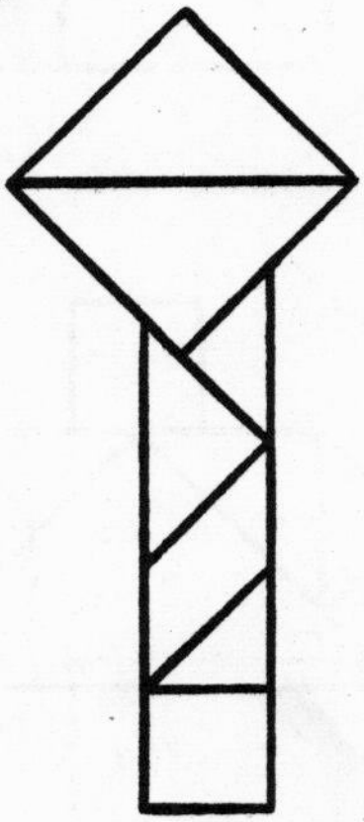

287 Mallet

288 Penknife

TANGRAM PARADOXES

For a resolution of the paradox in Tangrams 289 and 290 I cannot do better than repeat Dudeney's explanation: "It will be noticed that in both cases the head, hat and arm are precisely alike, and the width at the base of the body is the same. But this body contains four pieces in the first case, and in the second design only three. The first is larger than the second by exactly that narrow strip indicated by the dotted line between A and B. This strip is therefore exactly equal in area to the piece forming the foot in the other design, though when thus distributed along the side of the body the increased dimension is not easily apparent to the eye."

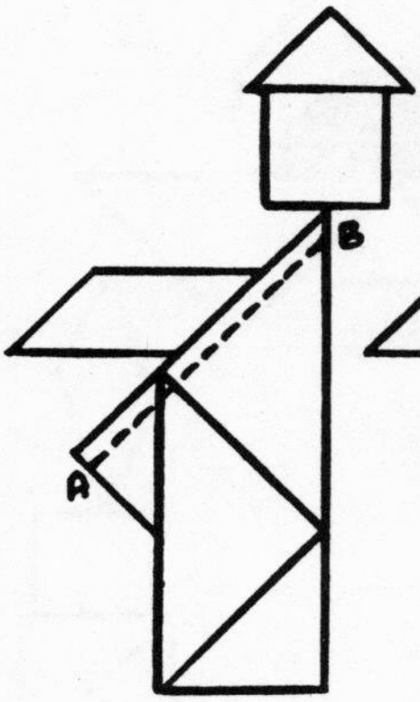

289

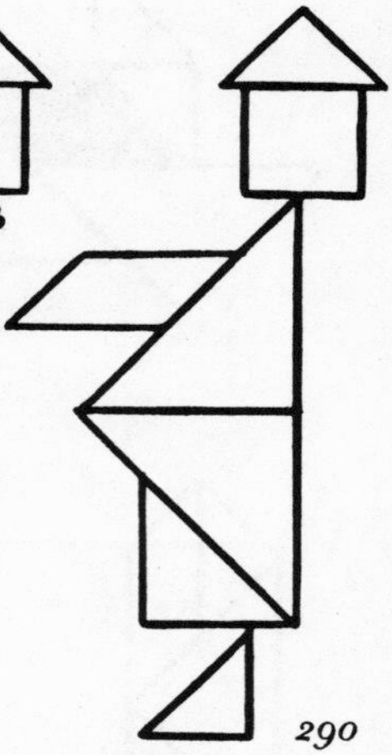

290

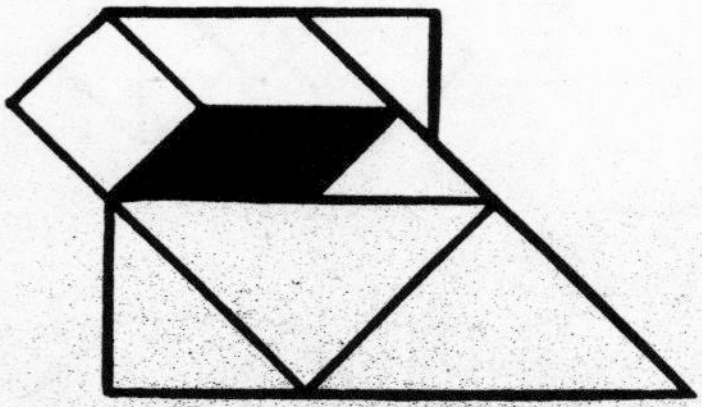

291

292

293

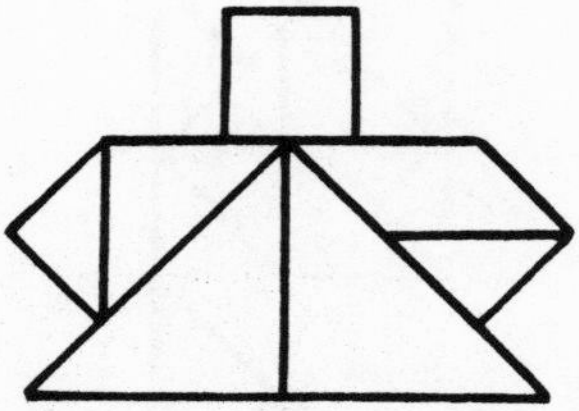

294

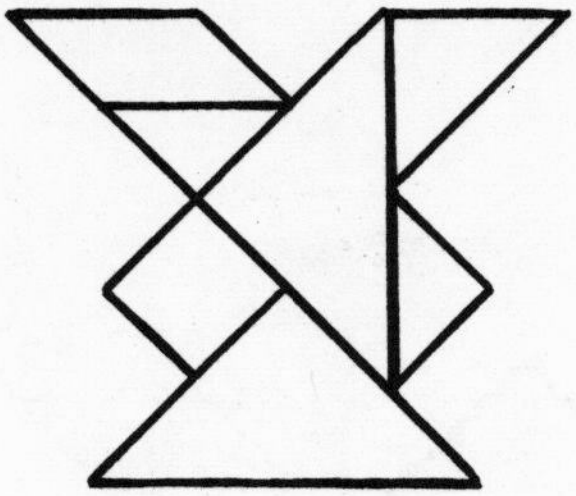

295

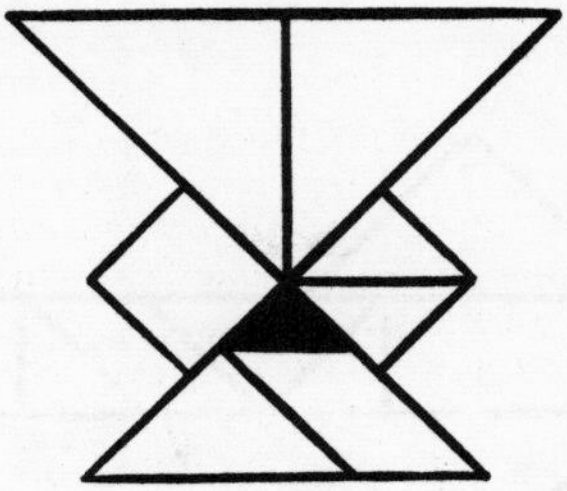

296

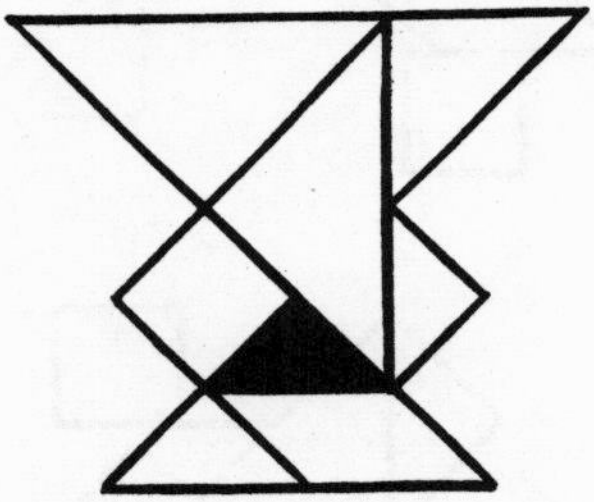

297

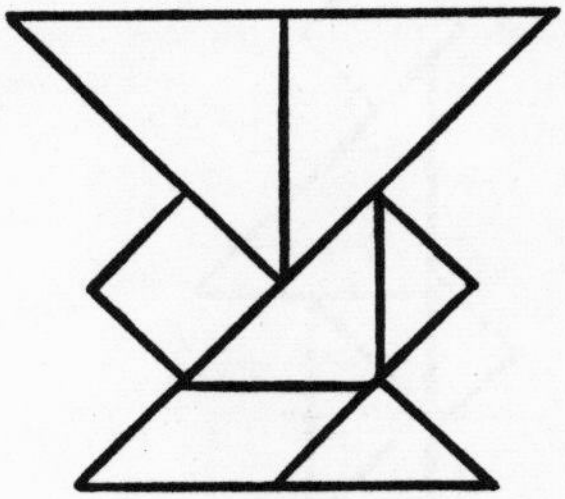

298

299

300

ILLUSIONS

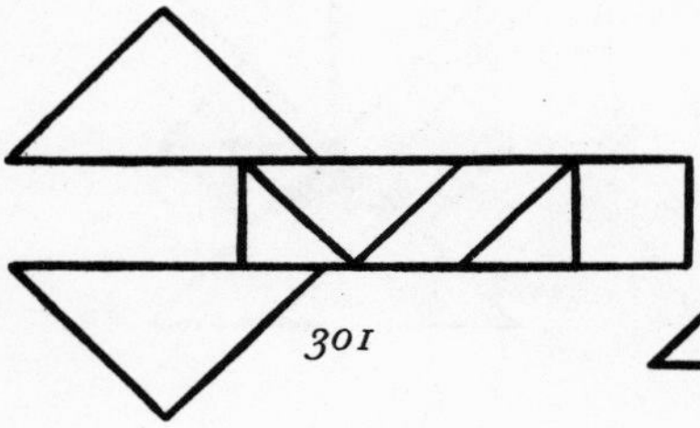

301

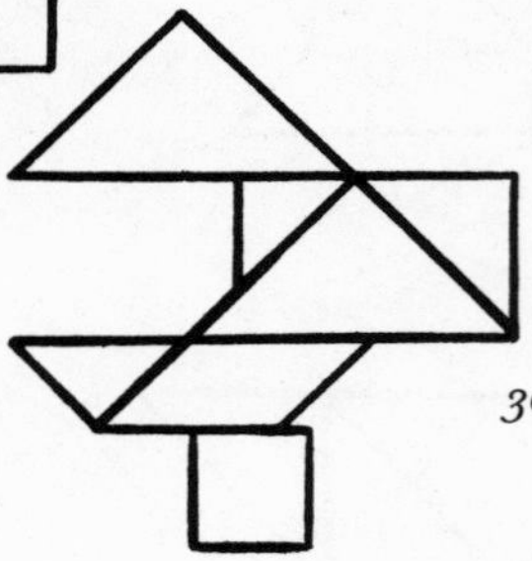

302

303

304

305

306

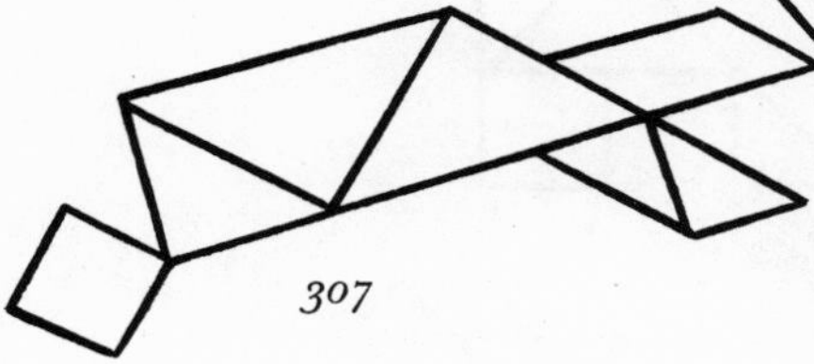

307

308

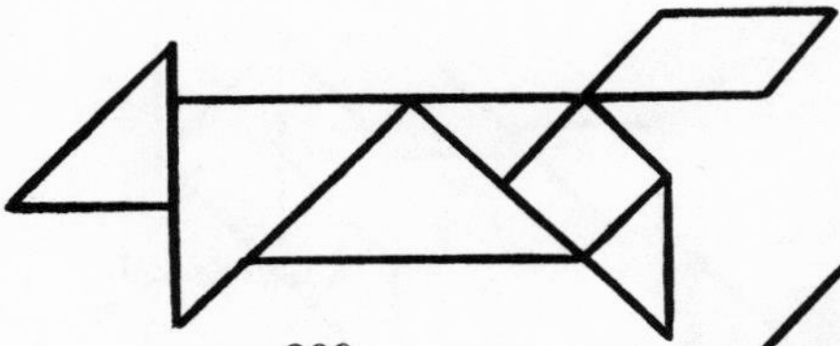

309

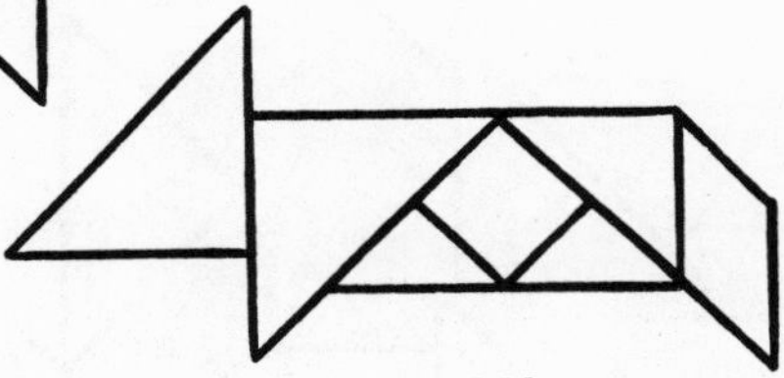

310

311

312 House

313 Cyclist

314 Dog

315 Lion

316 Car (ancient vintage)

317 Sports car

318 Coach

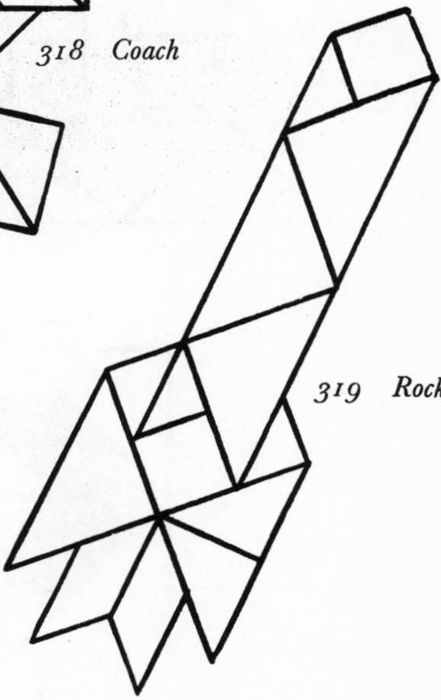

319 Rocket

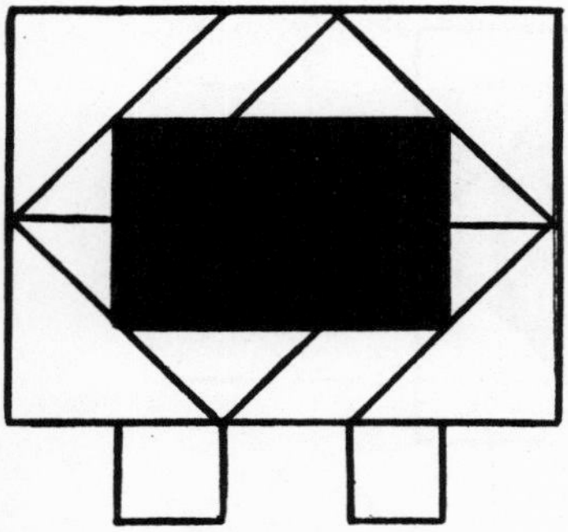

320 Television set

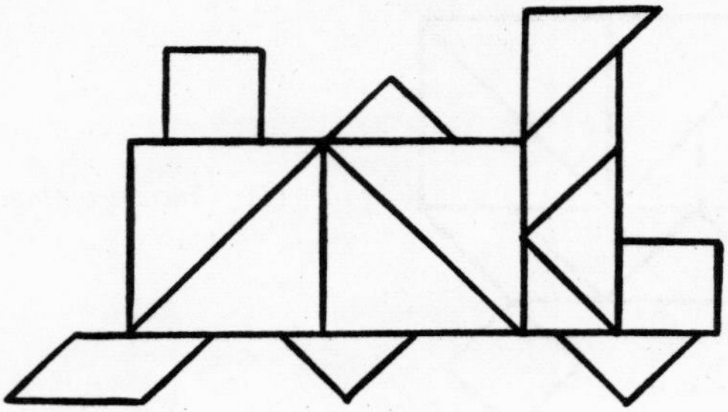

321 *Locomotive*

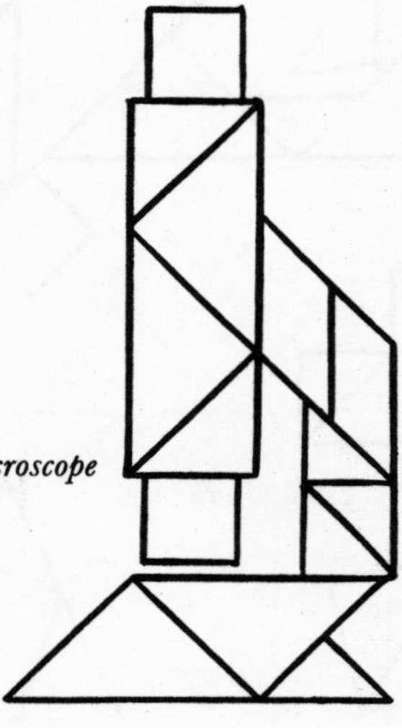

322 *Microscope*

323 *Telescope*

324 *Alarm clock*

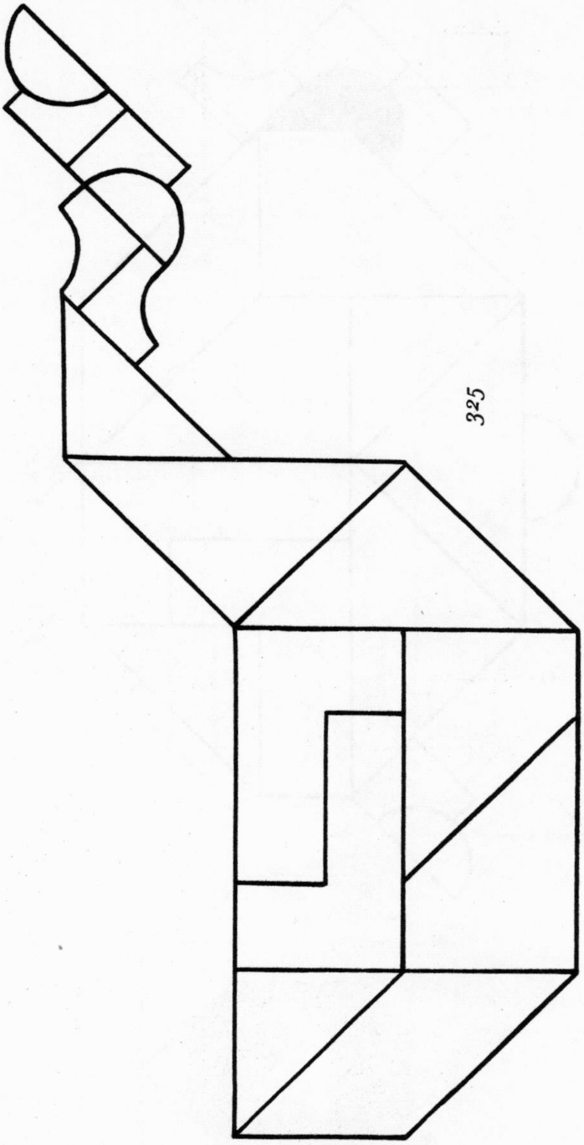

325

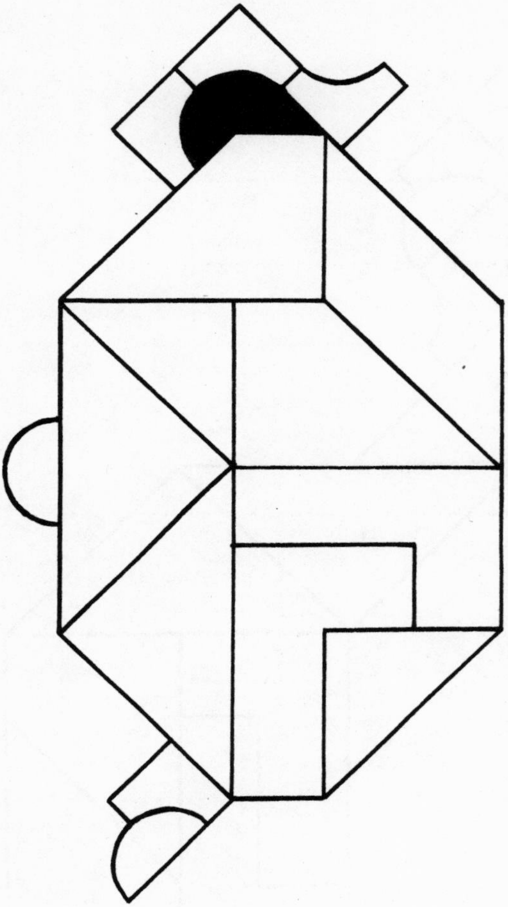

326

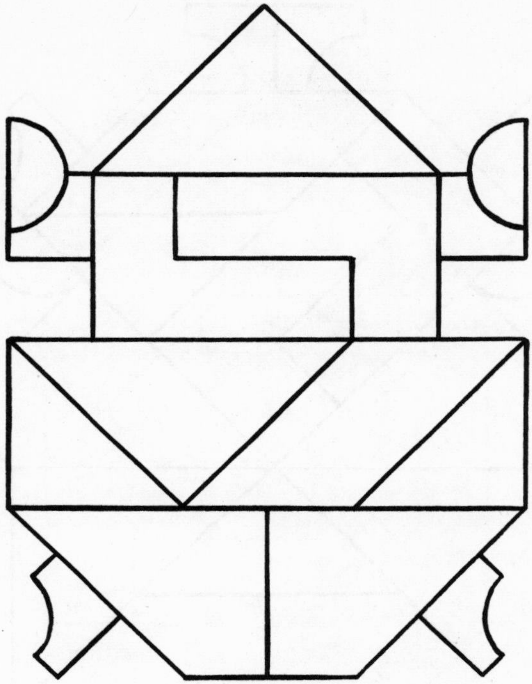

327

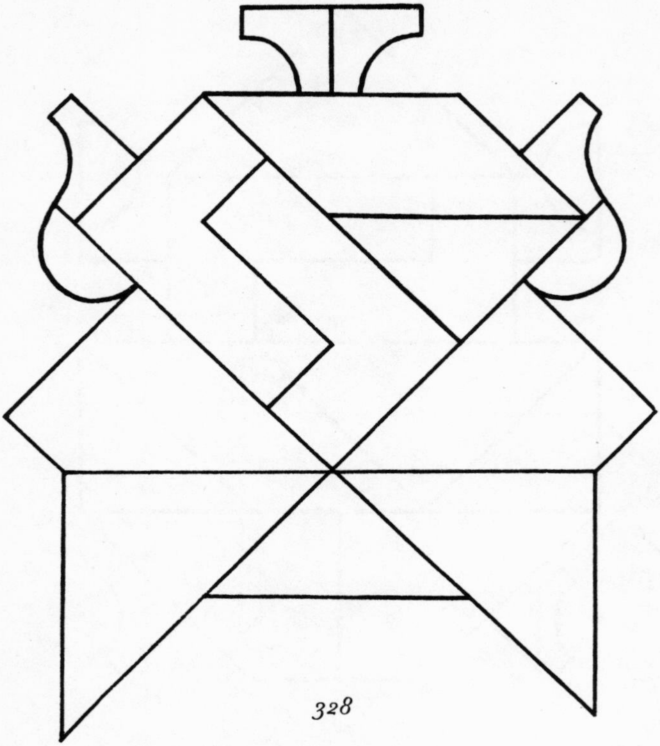

328

329

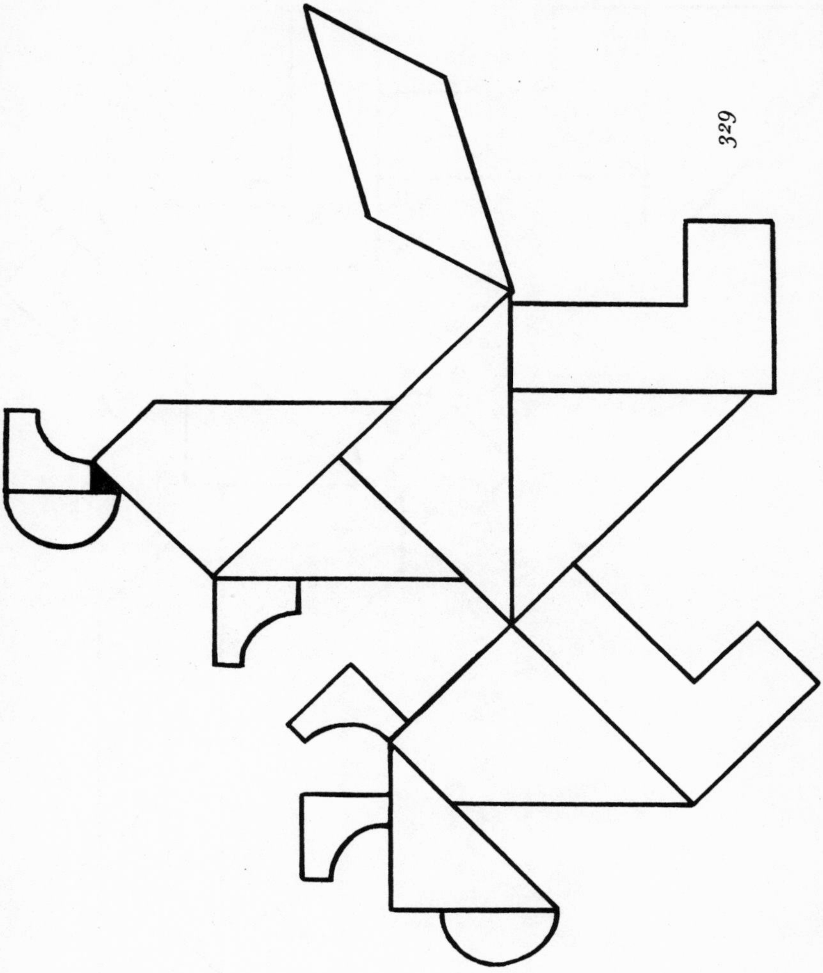

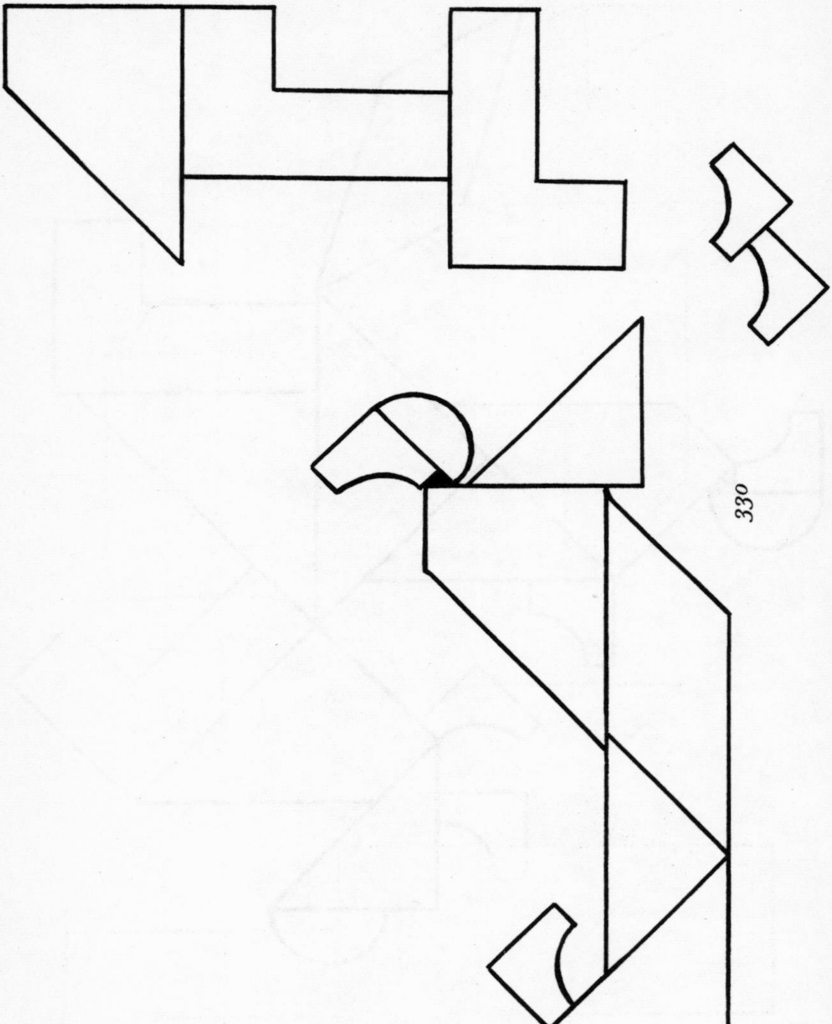

330

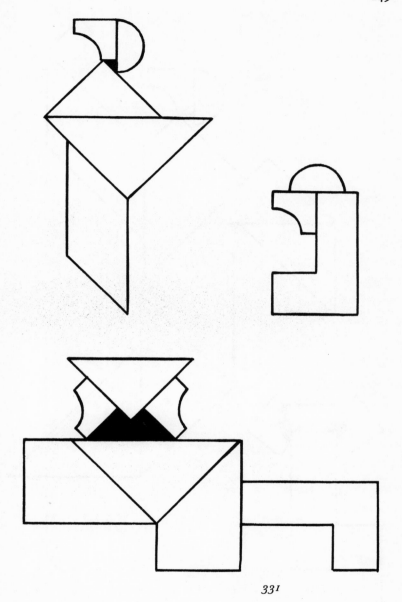

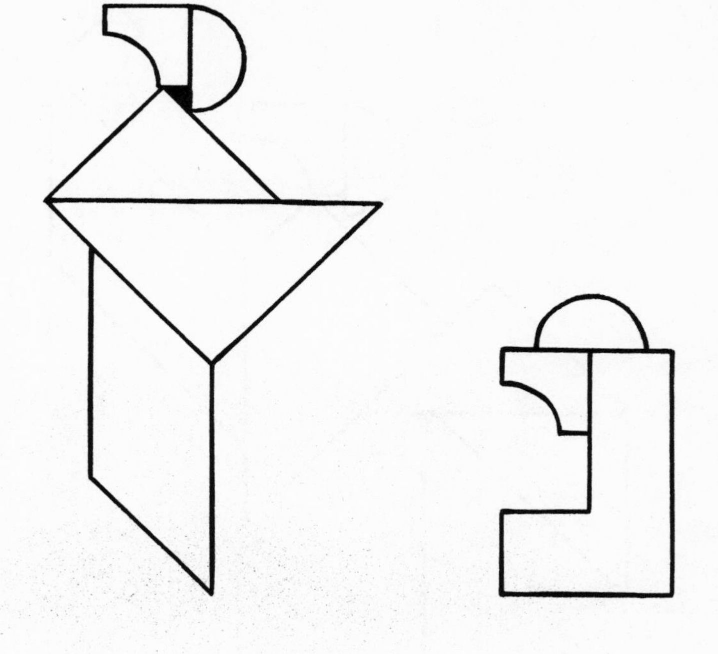

331

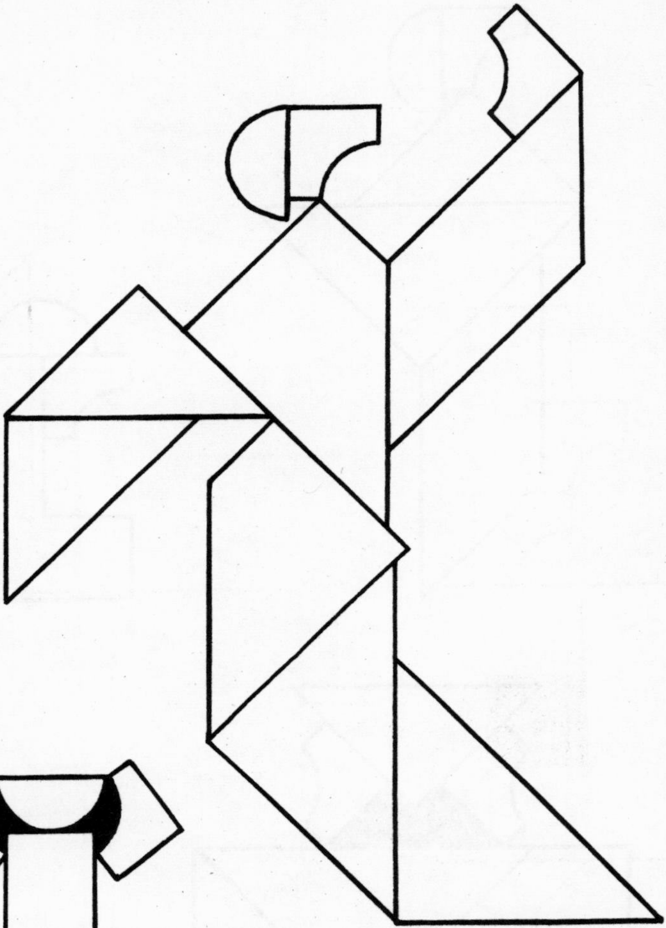

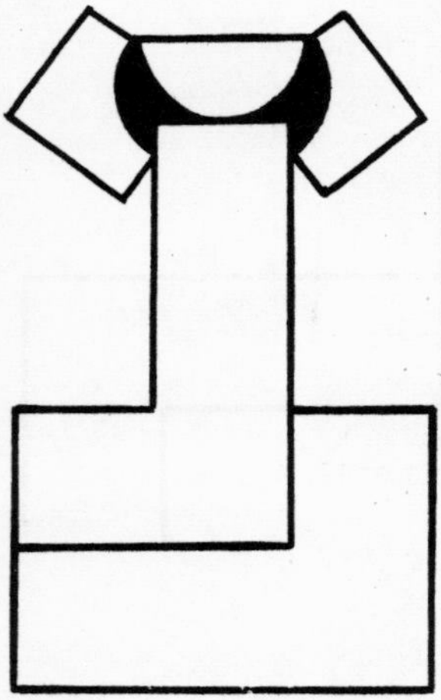

332

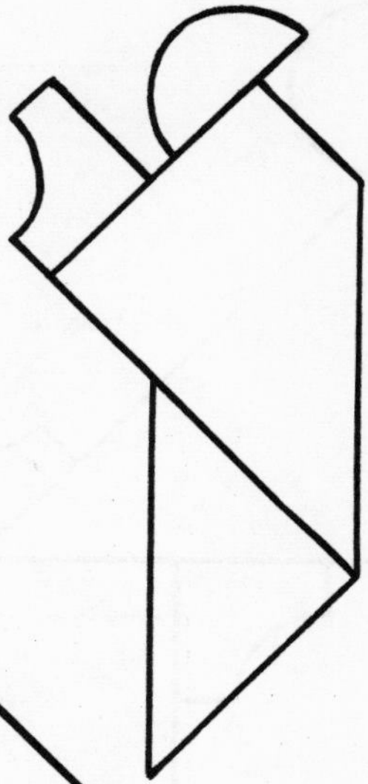

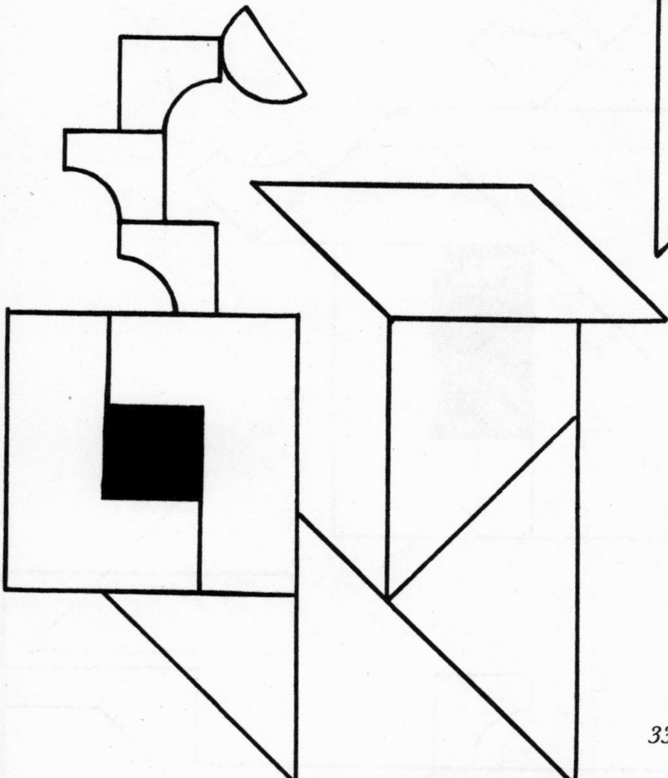

333

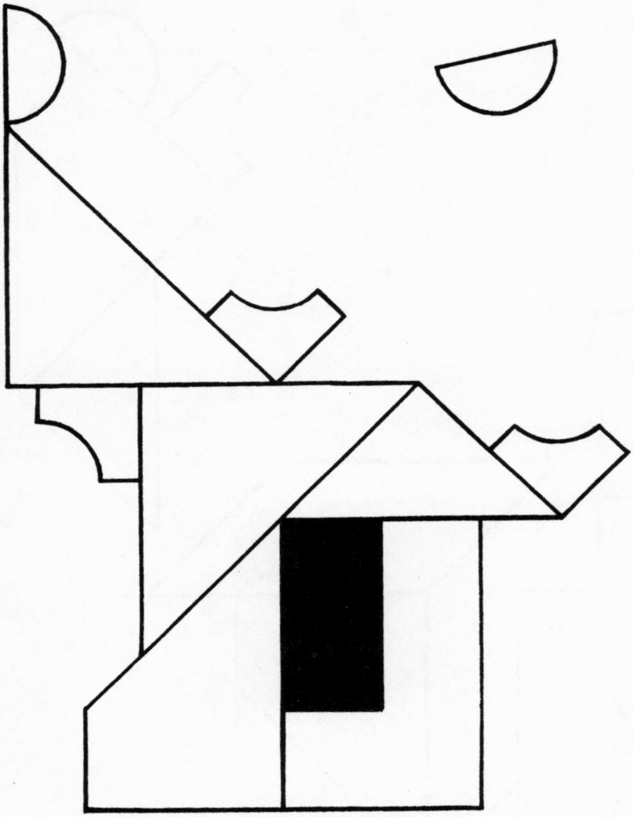

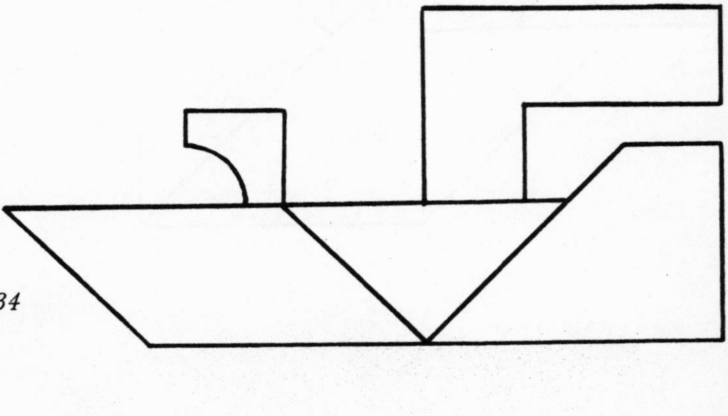

334

CATALOGUE OF DOVER BOOKS

Books Explaining Science and Mathematics

WHAT IS SCIENCE?, N. Campbell. The role of experiment and measurement, the function of mathematics, the nature of scientific laws, the difference between laws and theories, the limitations of science, and many similarly provocative topics are treated clearly and without technicalities by an eminent scientist. "Still an excellent introduction to scientific philosophy," H. Margenau in PHYSICS TODAY. "A first-rate primer . . . deserves a wide audience," SCIENTIFIC AMERICAN. 192pp. 5⅜ x 8. S43 Paperbound **$1.25**

THE NATURE OF PHYSICAL THEORY, P. W. Bridgman. A Nobel Laureate's clear, non-technical lectures on difficulties and paradoxes connected with frontier research on the physical sciences. Concerned with such central concepts as thought, logic, mathematics, relativity, probability, wave mechanics, etc. he analyzes the contributions of such men as Newton, Einstein, Bohr, Heisenberg, and many others. "Lucid and entertaining . . . recommended to anyone who wants to get some insight into current philosophies of science," THE NEW PHILOSOPHY. Index. xi + 138pp. 5⅜ x 8. S33 Paperbound **$1.25**

EXPERIMENT AND THEORY IN PHYSICS, Max Born. A Nobel Laureate examines the nature of experiment and theory in theoretical physics and analyzes the advances made by the great physicists of our day: Heisenberg, Einstein, Bohr, Planck, Dirac, and others. The actual process of creation is detailed step-by-step by one who participated. A fine examination of the scientific method at work. 44pp. 5⅜ x 8. S308 Paperbound **75¢**

THE PSYCHOLOGY OF INVENTION IN THE MATHEMATICAL FIELD, J. Hadamard. The reports of such men as Descartes, Pascal, Einstein, Poincaré, and others are considered in this investigation of the method of idea-creation in mathematics and other sciences and the thinking process in general. How do ideas originate? What is the role of the unconscious? What is Poincaré's forgetting hypothesis? are some of the fascinating questions treated. A penetrating analysis of Einstein's thought processes concludes the book. xiii + 145pp. 5⅜ x 8. T107 Paperbound **$1.25**

THE NATURE OF LIGHT AND COLOUR IN THE OPEN AIR, M. Minnaert. Why are shadows sometimes blue, sometimes green, or other colors depending on the light and surroundings? What causes mirages? Why do multiple suns and moons appear in the sky? Professor Minnaert explains these unusual phenomena and hundreds of others in simple, easy-to-understand terms based on optical laws and the properties of light and color. No mathematics is required but artists, scientists, students, and everyone fascinated by these "tricks" of nature will find thousands of useful and amazing pieces of information. Hundreds of observational experiments are suggested which require no special equipment. 200 illustrations; 42 photos. xvi + 362pp. 5⅜ x 8. T196 Paperbound **$2.00**

THE UNIVERSE OF LIGHT, W. Bragg. Sir William Bragg, Nobel Laureate and great modern physicist, is also well known for his powers of clear exposition. Here he analyzes all aspects of light for the layman: lenses, reflection, refraction, the optics of vision, x-rays, the photoelectric effect, etc. He tells you what causes the color of spectra, rainbows, and soap bubbles, how magic mirrors work, and much more. Dozens of simple experiments are described. Preface. Index. 199 line drawings and photographs, including 2 full-page color plates. x + 283pp. 5⅜ x 8. T538 Paperbound **$1.85**

SOAP-BUBBLES: THEIR COLOURS AND THE FORCES THAT MOULD THEM, C. V. Boys. For continuing popularity and validity as scientific primer, few books can match this volume of easily-followed experiments, explanations. Lucid exposition of complexities of liquid films, surface tension and related phenomena, bubbles' reaction to heat, motion, music, magnetic fields. Experiments with capillary attraction, soap bubbles on frames, composite bubbles, liquid cylinders and jets, bubbles other than soap, etc. Wonderful introduction to scientific method, natural laws that have many ramifications in areas of modern physics. Only complete edition in print. New Introduction by S. Z. Lewin, New York University. 83 illustrations; 1 full-page color plate. xii + 190pp. 5⅜ x 8½. T542 Paperbound **95¢**

CATALOGUE OF DOVER BOOKS

THE STORY OF X-RAYS FROM RONTGEN TO ISOTOPES, A. R. Bleich, M.D. This book, by a member of the American College of Radiology, gives the scientific explanation of x-rays, their applications in medicine, industry and art, and their danger (and that of atmospheric radiation) to the individual and the species. You learn how radiation therapy is applied against cancer, how x-rays diagnose heart disease and other ailments, how they are used to examine mummies for information on diseases of early societies, and industrial materials for hidden weaknesses. 54 illustrations show x-rays of flowers, bones, stomach, gears with flaws, etc. 1st publication. Index. xix + 186pp. 5⅜ x 8. T622 Paperbound **$1.50**

SPINNING TOPS AND GYROSCOPIC MOTION, John Perry. A classic elementary text of the dynamics of rotation — the behavior and use of rotating bodies such as gyroscopes and tops. In simple, everyday English you are shown how quasi-rigidity is induced in discs of paper, smoke rings, chains, etc., by rapid motions; why a gyrostat falls and why a top rises; precession; how the earth's motion affects climate; and many other phenomena. Appendix on practical use of gyroscopes. 62 figures. 128pp. 5⅜ x 8. T416 Paperbound **$1.25**

SNOW CRYSTALS, W. A. Bentley, M. J. Humphreys. For almost 50 years W. A. Bentley photographed snow flakes in his laboratory in Jericho, Vermont; in 1931 the American Meteorological Society gathered together the best of his work, some 2400 photographs of snow flakes, plus a few ice flowers, windowpane frosts, dew, frozen rain, and other ice formations. Pictures were selected for beauty and scientific value. A very valuable work to anyone in meteorology, cryology; most interesting to layman; extremely useful for artist who wants beautiful, crystalline designs. All copyright free. Unabridged reprint of 1931 edition. 2453 illustrations. 227pp. 8 x 10½. T287 Paperbound **$3.00**

A DOVER SCIENCE SAMPLER, edited by George Barkin. A collection of brief, non-technical passages from 44 Dover Books Explaining Science for the enjoyment of the science-minded browser. Includes work of Bertrand Russell, Poincaré, Laplace, Max Born, Galileo, Newton; material on physics, mathematics, metallurgy, anatomy, astronomy, chemistry, etc. You will be fascinated by Martin Gardner's analysis of the sincere pseudo-scientist, Moritz's account of Newton's absentmindedness, Bernard's examples of human vivisection, etc. Illustrations from the Diderot Pictorial Encyclopedia and De Re Metallica. 64 pages. **FREE**

THE STORY OF ATOMIC THEORY AND ATOMIC ENERGY, J. G. Feinberg. A broader approach to subject of nuclear energy and its cultural implications than any other similar source. Very readable, informal, completely non-technical text. Begins with first atomic theory, 600 B.C. and carries you through the work of Mendelejeff, Röntgen, Madame Curie, to Einstein's equation and the A-bomb. New chapter goes through thermonuclear fission, binding energy, other events up to 1959. Radioactive decay and radiation hazards, future benefits, work of Bohr, moderns, hundreds more topics. "Deserves special mention . . . not only authoritative but thoroughly popular in the best sense of the word," Saturday Review. Formerly, "The Atom Story." Expanded with new chapter. Three appendixes. Index. 34 illustrations. vii + 243pp. 5⅜ x 8. T625 Paperbound **$1.60**

THE STRANGE STORY OF THE QUANTUM, AN ACCOUNT FOR THE GENERAL READER OF THE GROWTH OF IDEAS UNDERLYING OUR PRESENT ATOMIC KNOWLEDGE, B. Hoffmann. Presents lucidly and expertly, with barest amount of mathematics, the problems and theories which led to modern quantum physics. Dr. Hoffmann begins with the closing years of the 19th century, when certain trifling discrepancies were noticed, and with illuminating analogies and examples takes you through the brilliant concepts of Planck, Einstein, Pauli, Broglie, Bohr, Schroedinger, Heisenberg, Dirac, Sommerfeld, Feynman, etc. This edition includes a new, long postscript carrying the story through 1958. "Of the books attempting an account of the history and contents of our modern atomic physics which have come to my attention, this is the best," H. Margenau, Yale University, in "American Journal of Physics." 32 tables and line illustrations. Index. 275pp. 5⅜ x 8. T518 Paperbound **$1.50**

SPACE AND TIME, E. Borel. Written by a versatile mathematician of world renown with his customary lucidity and precision, this introduction to relativity for the layman presents scores of examples, analogies, and illustrations that open up new ways of thinking about space and time. It covers abstract geometry and geographical maps, continuity and topology, the propagation of light, the special theory of relativity, the general theory of relativity, theoretical researches, and much more. Mathematical notes. 2 Indexes. 4 Appendices. 15 figures. xvi + 243pp. 5⅜ x 8. T592 Paperbound **$1.75**

FROM EUCLID TO EDDINGTON: A STUDY OF THE CONCEPTIONS OF THE EXTERNAL WORLD, Sir Edmund Whittaker. A foremost British scientist traces the development of theories of natural philosophy from the western rediscovery of Euclid to Eddington, Einstein, Dirac, etc. The inadequacy of classical physics is contrasted with present day attempts to understand the physical world through relativity, non-Euclidean geometry, space curvature, wave mechanics, etc. 5 major divisions of examination: Space; Time and Movement; the Concepts of Classical Physics; the Concepts of Quantum Mechanics; the Eddington Universe 212pp. 5⅜ x 8. T491 Paperbound **$1.35**

Nature, Biology,

NATURE RECREATION: Group Guidance for the Out-of-doors, William Gould Vinal. Intended for both the uninitiated nature instructor and the education student on the college level, this complete "how-to" program surveys the entire area of nature education for the young. Philosophy of nature recreation; requirements, responsibilities, important information for group leaders; nature games; suggested group projects; conducting meetings and getting discussions started; etc. Scores of immediately applicable teaching aids, plus completely updated sources of information, pamphlets, field guides, recordings, etc. Bibliography. 74 photographs. + 310pp. 5⅜ x 8½. T1015 Paperbound **$1.75**

HOW TO KNOW THE WILD FLOWERS, Mrs. William Starr Dana. Classic nature book that has introduced thousands to wonders of American wild flowers. Color-season principle of organization is easy to use, even by those with no botanical training, and the genial, refreshing discussions of history, folklore, uses of over 1,000 native and escape flowers, foliage plants are informative as well as fun to read. Over 170 full-page plates, collected from several editions, may be colored in to make permanent records of finds. Revised to conform with 1950 edition of Gray's Manual of Botany. xlii + 438pp. 5⅜ x 8½. T332 Paperbound **$2.00**

HOW TO KNOW THE FERNS, F. T. Parsons. Ferns, among our most lovely native plants, are all too little known. This classic of nature lore will enable the layman to identify almost any American fern he may come across. After an introduction on the structure and life of ferns, the 57 most important ferns are fully pictured and described (arranged upon a simple identification key). Index of Latin and English names. 61 illustrations and 42 full-page plates. xiv + 215pp. 5⅜ x 8. T740 Paperbound **$1.35**

MANUAL OF THE TREES OF NORTH AMERICA, Charles Sprague Sargent. Still unsurpassed as most comprehensive, reliable study of North American tree characteristics, precise locations and distribution. By dean of American dendrologists. Every tree native to U.S., Canada, Alaska, 185 genera, 717 species, described in detail—leaves, flowers, fruit, winterbuds, bark, wood, growth habits etc. plus discussion of varieties and local variants, immaturity variations. Over 100 keys, including unusual 11-page analytical key to genera, aid in identification. 783 clear illustrations of flowers, fruit, leaves. An unmatched permanent reference work for all nature lovers. Second enlarged (1926) edition. Synopsis of families. Analytical key to genera. Glossary of technical terms. Index. 783 illustrations, 1 map. Two volumes. Total of 982pp. 5⅜ x 8.
T277 Vol. I Paperbound **$2.25**
T278 Vol. II Paperbound **$2.25**
The set **$4.50**

TREES OF THE EASTERN AND CENTRAL UNITED STATES AND CANADA, W. M. Harlow. A revised edition of a standard middle-level guide to native trees and important escapes. More than 140 trees are described in detail, and illustrated with more than 600 drawings and photographs. Supplementary keys will enable the careful reader to identify almost any tree he might encounter. xiii + 288pp. 5⅜ x 8. T395 Paperbound **$1.35**

GUIDE TO SOUTHERN TREES, Ellwood S. Harrar and J. George Harrar. All the essential information about trees indigenous to the South, in an extremely handy format. Introductory essay on methods of tree classification and study, nomenclature, chief divisions of Southern trees, etc. Approximately 100 keys and synopses allow for swift, accurate identification of trees. Numerous excellent illustrations, non-technical text make this a useful book for teachers of biology or natural science, nature lovers, amateur naturalists. Revised 1962 edition. Index. Bibliography. Glossary of technical terms. 920 illustrations; 201 full-page plates. ix + 709pp. 4⅝ x 6⅜. T945 Paperbound **$2.35**

FRUIT KEY AND TWIG KEY TO TREES AND SHRUBS, W. M. Harlow. Bound together in one volume for the first time, these handy and accurate keys to fruit and twig identification are the only guides of their sort with photographs (up to 3 times natural size). "Fruit Key": Key to over 120 different deciduous and evergreen fruits. 139 photographs and 11 line drawings. Synoptic summary of fruit types. Bibliography. 2 Indexes (common and scientific names). "Twig Key": Key to over 160 different twigs and buds. 173 photographs. Glossary of technical terms. Bibliography. 2 Indexes (common and scientific names). Two volumes bound as one. Total of xvii + 126pp. 5⅝ x 8⅜. T511 Paperbound **$1.25**

INSECT LIFE AND INSECT NATURAL HISTORY, S. W. Frost. A work emphasizing habits, social life, and ecological relations of insects, rather than more academic aspects of classification and morphology. Prof. Frost's enthusiasm and knowledge are everywhere evident as he discusses insect associations and specialized habits like leaf-rolling, leaf-mining, and case-making, the gall insects, the boring insects, aquatic insects, etc. He examines all sorts of matters not usually covered in general works, such as: insects as human food, insect music and musicians, insect response to electric and radio waves, use of insects in art and literature. The admirably executed purpose of this book, which covers the middle ground between elementary treatment and scholarly monographs, is to excite the reader to observe for himself. Over 700 illustrations. Extensive bibliography. x + 524pp. 5⅜ x 8. T517 Paperbound **$2.50**

COMMON SPIDERS OF THE UNITED STATES, J. H. Emerton. Here is a nature hobby you can pursue right in your own cellar! Only non-technical, but thorough, reliable guide to spiders for the layman. Over 200 spiders from all parts of the country, arranged by scientific classification, are identified by shape and color, number of eyes, habitat and range, habits, etc. Full text, 501 line drawings and photographs, and valuable introduction explain webs, poisons, threads, capturing and preserving spiders, etc. Index. New synoptic key by S. W. Frost. xxiv + 225pp. 5⅜ x 8. T223 Paperbound **$1.45**

THE LIFE STORY OF THE FISH: HIS MANNERS AND MORALS, Brian Curtis. A comprehensive, non-technical survey of just about everything worth knowing about fish. Written for the aquarist, the angler, and the layman with an inquisitive mind, the text covers such topics as evolution, external covering and protective coloration, physics and physiology of vision, maintenance of equilibrium, function of the lateral line canal for auditory and temperature senses, nervous system, function of the air bladder, reproductive system and methods—courtship, mating, spawning, care of young—and many more. Also sections on game fish, the problems of conservation and a fascinating chapter on fish curiosities. "Clear, simple language . . . excellent judgment in choice of subjects . . . delightful sense of humor," New York Times. Revised (1949) edition. Index. Bibliography of 72 items. 6 full-page photographic plates. xii + 284pp. 5⅜ x 8. T929 Paperbound **$1.65**

BATS, Glover Morrill Allen. The most comprehensive study of bats as a life-form by the world's foremost authority. A thorough summary of just about everything known about this fascinating and mysterious flying mammal, including its unique location sense, hibernation and cycles, its habitats and distribution, its wing structure and flying habits, and its relationship to man in the long history of folklore and superstition. Written on a middle-level, the book can be profitably studied by a trained zoologist and thoroughly enjoyed by the layman. "An absorbing text with excellent illustrations. Bats should have more friends and fewer thoughtless detractors as a result of the publication of this volume," William Beebe, Books. Extensive bibliography. 57 photographs and illustrations. x + 368pp. 5⅜ x 8½. T984 Paperbound **$2.00**

BIRDS AND THEIR ATTRIBUTES, Glover Morrill Allen. A fine general introduction to birds as living organisms, especially valuable because of emphasis on structure, physiology, habits, behavior. Discusses relationship of bird to man, early attempts at scientific ornithology, feathers and coloration, skeletal structure including bills, legs and feet, wings. Also food habits, evolution and present distribution, feeding and nest-building, still unsolved questions of migrations and location sense, many more similar topics. Final chapter on classification, nomenclature. A good popular-level summary for the biologist; a first-rate introduction for the layman. Reprint of 1925 edition. References and index. 51 illustrations. viii + 338pp. 5⅜ x 8½. T957 Paperbound **$1.85**

LIFE HISTORIES OF NORTH AMERICAN BIRDS, Arthur Cleveland Bent. Bent's monumental series of books on North American birds, prepared and published under auspices of Smithsonian Institute, is the definitive coverage of the subject, the most-used single source of information. Now the entire set is to be made available by Dover in inexpensive editions. This encyclopedic collection of detailed, specific observations utilizes reports of hundreds of contemporary observers, writings of such naturalists as Audubon, Burroughs, William Brewster, as well as author's own extensive investigations. Contains literally everything known about life history of each bird considered: nesting, eggs, plumage, distribution and migration, voice, enemies, courtship, etc. These not over-technical works are musts for ornithologists, conservationists, amateur naturalists, anyone seriously interested in American birds.

BIRDS OF PREY. More than 100 subspecies of hawks, falcons, eagles, buzzards, condors and owls, from the common barn owl to the extinct caracara of Guadaloupe Island. 400 photographs. Two volume set. Index for each volume. Bibliographies of 403, 520 items. 197 full-page plates. Total of 907pp. 5⅜ x 8½. Vol. I T931 Paperbound **$2.50**
Vol. II T932 Paperbound **$2.50**

WILD FOWL. Ducks, geese, swans, and tree ducks—73 different subspecies. Two volume set. Index for each volume. Bibliographies of 124, 144 items. 106 full-page plates. Total of 685pp. 5⅜ x 8½. Vol. I T285 Paperbound **$2.50**
Vol. II T286 Paperbound **$2.50**

SHORE BIRDS. 81 varieties (sandpipers, woodcocks, plovers, snipes, phalaropes, curlews, oyster catchers, etc.). More than 200 photographs of eggs, nesting sites, adult and young of important species. Two volume set. Index for each volume. Bibliographies of 261, 188 items. 121 full-page plates. Total of 860pp. 5⅜ x 8½. Vol. I T933 Paperbound **$2.35**
Vol. II T934 Paperbound **$2.35**

THE LIFE OF PASTEUR, R. Vallery-Radot. 13th edition of this definitive biography, cited in Encyclopaedia Britannica. Authoritative, scholarly, well-documented with contemporary quotes, observations; gives complete picture of Pasteur's personal life; especially thorough presentation of scientific activities with silkworms, fermentation, hydrophobia, inoculation, etc. Introduction by Sir William Osler. Index. 505pp. 5⅜ x 8. T632 Paperbound **$2.00**

Puzzles, Mathematical Recreations

SYMBOLIC LOGIC and THE GAME OF LOGIC, Lewis Carroll. "Symbolic Logic" is not concerned with modern symbolic logic, but is instead a collection of over 380 problems posed with charm and imagination, using the syllogism, and a fascinating diagrammatic method of drawing conclusions. In "The Game of Logic" Carroll's whimsical imagination devises a logical game played with 2 diagrams and counters (included) to manipulate hundreds of tricky syllogisms. The final section, "Hit or Miss" is a lagniappe of 101 additional puzzles in the delightful Carroll manner. Until this reprint edition, both of these books were rarities costing up to $15 each. Symbolic Logic: Index. xxxi + 199pp. The Game of Logic: 96pp. 2 vols. bound as one. 5⅜ x 8. T492 Paperbound **$1.75**

PILLOW PROBLEMS and A TANGLED TALE, Lewis Carroll. One of the rarest of all Carroll's works, "Pillow Problems" contains 72 original math puzzles, all typically ingenious. Particularly fascinating are Carroll's answers which remain exactly as he thought them out, reflecting his actual mental process. The problems in "A Tangled Tale" are in story form, originally appearing as a monthly magazine serial. Carroll not only gives the solutions, but uses answers sent in by readers to discuss wrong approaches and misleading paths, and grades them for insight. Both of these books were rarities until this edition, "Pillow Problems" costing up to $25, and "A Tangled Tale" $15. Pillow Problems: Preface and Introduction by Lewis Carroll. xx + 109pp. A Tangled Tale: 6 illustrations. 152pp. Two vols. bound as one. 5⅜ x 8. T493 Paperbound **$1.50**

AMUSEMENTS IN MATHEMATICS, Henry Ernest Dudeney. The foremost British originator of mathematical puzzles is always intriguing, witty, and paradoxical in this classic, one of the largest collections of mathematical amusements. More than 430 puzzles, problems, and paradoxes. Mazes and games, problems on number manipulation, unicursal and other route problems, puzzles on measuring, weighing, packing, age, kinship, chessboards, joiners', crossing river, plane figure dissection, and many others. Solutions. More than 450 illustrations. vii + 258pp. 5⅜ x 8. T473 Paperbound **$1.25**

THE CANTERBURY PUZZLES, Henry Dudeney. Chaucer's pilgrims set one another problems in story form. Also Adventures of the Puzzle Club, the Strange Escape of the King's Jester, the Monks of Riddlewell, the Squire's Christmas Puzzle Party, and others. All puzzles are original, based on dissecting plane figures, arithmetic, algebra, elementary calculus and other branches of mathematics, and purely logical ingenuity. "The limit of ingenuity and intricacy," The Observer. Over 110 puzzles. Full Solutions. 150 illustrations. vii + 225pp. 5⅜ x 8. T474 Paperbound **$1.25**

MATHEMATICAL EXCURSIONS, H. A. Merrill. Even if you hardly remember your high school math, you'll enjoy the 90 stimulating problems contained in this book and you will come to understand a great many mathematical principles with surprisingly little effort. Many useful shortcuts and diversions not generally known are included: division by inspection, Russian peasant multiplication, memory systems for pi, building odd and even magic squares, square roots by geometry, dyadic systems, and many more. Solutions to difficult problems. 50 illustrations. 145pp. 5⅜ x 8. T350 Paperbound **$1.00**

MAGIC SQUARES AND CUBES, W. S. Andrews. Only book-length treatment in English, a thorough non-technical description and analysis. Here are nasik, overlapping, pandiagonal, serrated squares; magic circles, cubes, spheres, rhombuses. Try your hand at 4-dimensional magical figures! Much unusual folklore and tradition included. High school algebra is sufficient. 754 diagrams and illustrations. viii + 419pp. 5⅜ x 8. T658 Paperbound **$1.85**

CALIBAN'S PROBLEM BOOK: MATHEMATICAL, INFERENTIAL AND CRYPTOGRAPHIC PUZZLES, H. Phillips (Caliban), S. T. Shovelton, G. S. Marshall. 105 ingenious problems by the greatest living creator of puzzles based on logic and inference. Rigorous, modern, piquant; reflecting their author's unusual personality, these intermediate and advanced puzzles all involve the ability to reason clearly through complex situations; some call for mathematical knowledge, ranging from algebra to number theory. Solutions. xi + 180pp. 5⅜ x 8. T736 Paperbound **$1.25**

MATHEMATICAL PUZZLES FOR BEGINNERS AND ENTHUSIASTS, G. Mott-Smith. 188 mathematical puzzles based on algebra, dissection of plane figures, permutations, and probability, that will test and improve your powers of inference and interpretation. The Odic Force, The Spider's Cousin, Ellipse Drawing, theory and strategy of card and board games like tit-tat-toe, go moku, salvo, and many others. 100 pages of detailed mathematical explanations. Appendix of primes, square roots, etc. 135 illustrations. 2nd revised edition. 248pp. 5⅜ x 8. T198 Paperbound **$1.00**

MATHEMAGIC, MAGIC PUZZLES, AND GAMES WITH NUMBERS, R. V. Heath. More than 60 new puzzles and stunts based on the properties of numbers. Easy techniques for multiplying large numbers mentally, revealing hidden numbers magically, finding the date of any day in any year, and dozens more. Over 30 pages devoted to magic squares, triangles, cubes, circles, etc. Edited by J. S. Meyer. 76 illustrations. 128pp. 5⅜ x 8. T110 Paperbound **$1.00**

THE BOOK OF MODERN PUZZLES, G. L. Kaufman. A completely new series of puzzles as fascinating as crossword and deduction puzzles but based upon different principles and techniques. Simple 2-minute teasers, word labyrinths, design and pattern puzzles, logic and observation puzzles — over 150 braincrackers. Answers to all problems. 116 illustrations. 192pp. 5⅜ x 8.
T143 Paperbound **$1.00**

NEW WORD PUZZLES, G. L. Kaufman. 100 ENTIRELY NEW puzzles based on words and their combinations that will delight crossword puzzle, Scrabble and Jotto fans. Chess words, based on the moves of the chess king; design-onyms, symmetrical designs made of synonyms; rhymed double-crostics; syllable sentences; addle letter anagrams; alphagrams; linkograms; and many others all brand new. Full solutions. Space to work problems. 196 figures. vi + 122pp. 5⅜ x 8.
T344 Paperbound **$1.00**

MAZES AND LABYRINTHS: A BOOK OF PUZZLES, W. Shepherd. Mazes, formerly associated with mystery and ritual, are still among the most intriguing of intellectual puzzles. This is a novel and different collection of 50 amusements that embody the principle of the maze: mazes in the classical tradition; 3-dimensional, ribbon, and Möbius-strip mazes; hidden messages; spatial arrangements; etc.—almost all built on amusing story situations. 84 illustrations. Essay on maze psychology. Solutions. xv + 122pp. 5⅜ x 8.
T731 Paperbound **$1.00**

MAGIC TRICKS & CARD TRICKS, W. Jonson. Two books bound as one. 52 tricks with cards, 37 tricks with coins, bills, eggs, smoke, ribbons, slates, etc. Details on presentation, misdirection, and routining will help you master such famous tricks as the Changing Card, Card in the Pocket, Four Aces, Coin Through the Hand, Bill in the Egg, Afghan Bands, and over 75 others. If you follow the lucid exposition and key diagrams carefully, you will finish these two books with an astonishing mastery of magic. 106 figures. 224pp. 5⅜ x 8.
T909 Paperbound **$1.00**

PANORAMA OF MAGIC, Milbourne Christopher. A profusely illustrated history of stage magic, a unique selection of prints and engravings from the author's private collection of magic memorabilia, the largest of its kind. Apparatus, stage settings and costumes; ingenious ads distributed by the performers and satiric broadsides passed around in the streets ridiculing pompous showmen; programs; decorative souvenirs. The lively text, by one of America's foremost professional magicians, is full of anecdotes about almost legendary wizards: Dede, the Egyptian; Philadelphia, the wonder-worker; Robert-Houdin, "the father of modern magic;" Harry Houdini; scores more. Altogether a pleasure package for anyone interested in magic, stage setting and design, ethnology, psychology, or simply in unusual people. A Dover original. 295 illustrations; 8 in full color. Index. viii + 216pp. 8⅜ x 11¼.
T774 Paperbound **$2.25**

HOUDINI ON MAGIC, Harry Houdini. One of the greatest magicians of modern times explains his most prized secrets. How locks are picked, with illustrated picks and skeleton keys; how a girl is sawed into twins; how to walk through a brick wall — Houdini's explanations of 44 stage tricks with many diagrams. Also included is a fascinating discussion of great magicians of the past and the story of his fight against fraudulent mediums and spiritualists. Edited by W.B. Gibson and M.N. Young. Bibliography. 155 figures, photos. xv + 280pp. 5⅜ x 8.
T384 Paperbound **$1.35**

MATHEMATICS, MAGIC AND MYSTERY, Martin Gardner. Why do card tricks work? How do magicians perform astonishing mathematical feats? How is stage mind-reading possible? This is the first book length study explaining the application of probability, set theory, theory of numbers, topology, etc., to achieve many startling tricks. Non-technical, accurate, detailed! 115 sections discuss tricks with cards, dice, coins, knots, geometrical vanishing illusions, how a Curry square "demonstrates" that the sum of the parts may be greater than the whole, and dozens of others. No sleight of hand necessary! 135 illustrations. xii + 174pp. 5⅜ x 8.
T335 Paperbound **$1.00**

EASY-TO-DO ENTERTAINMENTS AND DIVERSIONS WITH COINS, CARDS, STRING, PAPER AND MATCHES, R. M. Abraham. Over 300 tricks, games and puzzles will provide young readers with absorbing fun. Sections on card games; paper-folding; tricks with coins, matches and pieces of string; games for the agile; toy-making from common household objects; mathematical recreations; and 50 miscellaneous pastimes. Anyone in charge of groups of youngsters, including hard-pressed parents, and in need of suggestions on how to keep children sensibly amused and quietly content will find this book indispensable. Clear, simple text, copious number of delightful line drawings and illustrative diagrams. Originally titled "Winter Nights Entertainments." Introduction by Lord Baden Powell. 329 illustrations. v + 186pp. 5⅜ x 8½.
T921 Paperbound **$1.00**

STRING FIGURES AND HOW TO MAKE THEM, Caroline Furness Jayne. 107 string figures plus variations selected from the best primitive and modern examples developed by Navajo, Apache, pygmies of Africa, Eskimo, in Europe, Australia, China, etc. The most readily understandable, easy-to-follow book in English on perennially popular recreation. Crystal-clear exposition; step-by-step diagrams. Everyone from kindergarten children to adults looking for unusual diversion will be endlessly amused. Index. Bibliography. Introduction by A. C. Haddon. 17 full-page plates. 960 illustrations. xxiii + 401pp. 5⅜ x 8½.
T152 Paperbound **$2.00**

Entertainments, Humor

ODDITIES AND CURIOSITIES OF WORDS AND LITERATURE, C. Bombaugh, edited by M. Gardner. The largest collection of idiosyncratic prose and poetry techniques in English, a legendary work in the curious and amusing bypaths of literary recreations and the play technique in literature—so important in modern works. Contains alphabetic poetry, acrostics, palindromes, scissors verse, centos, emblematic poetry, famous literary puns, hoaxes, notorious slips of the press, hilarious mistranslations, and much more. Revised and enlarged with modern material by Martin Gardner. 368pp. 5⅜ x 8. T759 Paperbound **$1.75**

A NONSENSE ANTHOLOGY, collected by **Carolyn Wells.** 245 of the best nonsense verses ever written, including nonsense puns, absurd arguments, mock epics and sagas, nonsense ballads, odes, "sick" verses, dog-Latin verses, French nonsense verses, songs. By Edward Lear, Lewis Carroll, Gelett Burgess, W. S. Gilbert, Hilaire Belloc, Peter Newell, Oliver Herford, etc., 83 writers in all plus over four score anonymous nonsense verses. A special section of limericks, plus famous nonsense such as Carroll's "Jabberwocky" and Lear's "The Jumblies" and much excellent verse virtually impossible to locate elsewhere. For 50 years considered the best anthology available. Index of first lines specially prepared for this edition. Introduction by Carolyn Wells. 3 indexes: Title, Author, First lines. xxxiii + 279pp. T499 Paperbound **$1.35**

THE BAD CHILD'S BOOK OF BEASTS, MORE BEASTS FOR WORSE CHILDREN, and A MORAL ALPHABET, H. Belloc. Hardly an anthology of humorous verse has appeared in the last 50 years without at least a couple of these famous nonsense verses. But one must see the entire volumes—with all the delightful original illustrations by Sir Basil Blackwood—to appreciate fully Belloc's charming and witty verses that play so subacidly on the platitudes of life and morals that beset his day—and ours. A great humor classic. Three books in one. Total of 157pp. 5⅜ x 8. T749 Paperbound **$1.00**

THE DEVIL'S DICTIONARY, Ambrose Bierce. Sardonic and irreverent barbs puncturing the pomposities and absurdities of American politics, business, religion, literature, and arts, by the country's greatest satirist in the classic tradition. Epigrammatic as Shaw, piercing as Swift, American as Mark Twain, Will Rogers, and Fred Allen, Bierce will always remain the favorite of a small coterie of enthusiasts, and of writers and speakers whom he supplies with "some of the most gorgeous witticisms of the English language" (H. L. Mencken). Over 1000 entries in alphabetical order. 144pp. 5⅜ x 8. T487 Paperbound **$1.00**

THE PURPLE COW AND OTHER NONSENSE, Gelett Burgess. The best of Burgess's early nonsense, selected from the first edition of the "Burgess Nonsense Book." Contains many of his most unusual and truly awe-inspiring pieces: 36 nonsense quatrains, the Poems of Patagonia, Alphabet of Famous Goops, and the other hilarious (and rare) adult nonsense that place him in the forefront of American humorists. All pieces are accompanied by the original Burgess illustrations. 123 illustrations. xiii + 113pp. 5⅜ x 8. T772 Paperbound **$1.00**

MY PIOUS FRIENDS AND DRUNKEN COMPANIONS and MORE PIOUS FRIENDS AND DRUNKEN COMPANIONS, Frank Shay. Folksingers, amateur and professional, and everyone who loves singing: here, available for the first time in 30 years, is this valued collection of 132 ballads, blues, vaudeville numbers, drinking songs, sea chanties, comedy songs. Songs of pre-Beatnik Bohemia; songs from all over America, England, France, Australia; the great songs of the Naughty Nineties and early twentieth-century America. Over a third with music. Woodcuts by John Held, Jr. convey perfectly the brash insouciance of an era of rollicking unabashed song. 12 illustrations by John Held, Jr. Two indexes (Titles and First lines and Choruses). Introductions by the author. Two volumes bound as one. Total of xvi + 235pp. 5⅜ x 8½. T946 Paperbound **$1.25**

HOW TO TELL THE BIRDS FROM THE FLOWERS, R. W. Wood. How not to confuse a carrot with a parrot, a grape with an ape, a puffin with nuffin. Delightful drawings, clever puns, absurd little poems point out far-fetched resemblances in nature. The author was a leading physicist. Introduction by Margaret Wood White. 106 illus. 60pp. 5⅜ x 8. T523 Paperbound **75¢**

PECK'S BAD BOY AND HIS PA, George W. Peck. The complete edition, containing both volumes, of one of the most widely read American humor books. The endless ingenious pranks played by bad boy "Hennery" on his pa and the grocery man, the outraged pomposity of Pa, the perpetual ridiculing of middle class institutions, are as entertaining today as they were in 1883. No pale sophistications or subtleties, but rather humor vigorous, raw, earthy, imaginative, and, as folk humor often is, sadistic. This peculiarly fascinating book is also valuable to historians and students of American culture as a portrait of an age. 100 original illustrations by True Williams. Introduction by E. F. Bleiler. 347pp. 5⅜ x 8. T497 Paperbound **$1.50**

THE HUMOROUS VERSE OF LEWIS CARROLL. Almost every poem Carroll ever wrote, the largest collection ever published, including much never published elsewhere: 150 parodies, burlesques, riddles, ballads, acrostics, etc., with 130 original illustrations by Tenniel, Carroll, and others. "Addicts will be grateful . . . there is nothing for the faithful to do but sit down and fall to the banquet," N. Y. Times. Index to first lines. xiv + 446pp. 5⅜ x 8.
T654 Paperbound **$2.00**

DIVERSIONS AND DIGRESSIONS OF LEWIS CARROLL. A major new treasure for Carroll fans! Rare privately published humor, fantasy, puzzles, and games by Carroll at his whimsical best, with a new vein of frank satire. Includes many new mathematical amusements and recreations, among them the fragmentary Part III of "Curiosa Mathematica." Contains "The Rectory Umbrella," "The New Belfry," "The Vision of the Three T's," and much more. New 32-page supplement of rare photographs taken by Carroll. x + 375pp. 5⅜ x 8.
T732 Paperbound **$2.00**

THE COMPLETE NONSENSE OF EDWARD LEAR. This is the only complete edition of this master of gentle madness available at a popular price. A BOOK OF NONSENSE, NONSENSE SONGS, MORE NONSENSE SONGS AND STORIES in their entirety with all the old favorites that have delighted children and adults for years. The Dong With A Luminous Nose, The Jumblies, The Owl and the Pussycat, and hundreds of other bits of wonderful nonsense. 214 limericks, 3 sets of Nonsense Botany, 5 Nonsense Alphabets, 546 drawings by Lear himself, and much more. 320pp. 5⅜ x 8. T167 Paperbound **$1.00**

THE MELANCHOLY LUTE, The Humorous Verse of Franklin P. Adams ("FPA"). The author's own selection of light verse, drawn from thirty years of FPA's column, "The Conning Tower," syndicated all over the English-speaking world. Witty, perceptive, literate, these ninety-six poems range from parodies of other poets, Millay, Longfellow, Edgar Guest, Kipling, Masefield, etc., and free and hilarious translations of Horace and other Latin poets, to satiric comments on fabled American institutions—the New York Subways, preposterous ads, suburbanites, sensational journalism, etc. They reveal with vigor and clarity the humor, integrity and restraint of a wise and gentle American satirist. Introduction by Robert Hutchinson. vi + 122pp. 5⅜ x 8½. T108 Paperbound **$1.00**

SINGULAR TRAVELS, CAMPAIGNS, AND ADVENTURES OF BARON MUNCHAUSEN, R. E. Raspe, with 90 illustrations by Gustave Doré. The first edition in over 150 years to reestablish the deeds of the Prince of Liars exactly as Raspe first recorded them in 1785—the genuine Baron Munchausen, one of the most popular personalities in English literature. Included also are the best of the many sequels, written by other hands. Introduction on Raspe by J. Carswell. Bibliography of early editions. xliv + 192pp. 5⅜ x 8. T698 Paperbound **$1.00**

THE WIT AND HUMOR OF OSCAR WILDE, ed. by Alvin Redman. Wilde at his most brilliant, in 1000 epigrams exposing weaknesses and hypocrisies of "civilized" society. Divided into 49 categories—sin, wealth, women, America, etc.—to aid writers, speakers. Includes excerpts from his trials, books, plays, criticism. Formerly "The Epigrams of Oscar Wilde." Introduction by Vyvyan Holland, Wilde's only living son. Introductory essay by editor. 260pp. 5⅜ x 8. T602 Paperbound **$1.00**

MAX AND MORITZ, Wilhelm Busch. Busch is one of the great humorists of all time, as well as the father of the modern comic strip. This volume, translated by H. A. Klein and other hands, contains the perennial favorite "Max and Moritz" (translated by C. T. Brooks), Plisch and Plum, Das Rabennest, Eispeter, and seven other whimsical, sardonic, jovial, diabolical cartoon and verse stories. Lively English translations parallel the original German. This work has delighted millions since it first appeared in the 19th century, and is guaranteed to please almost anyone. Edited by H. A. Klein, with an afterword. x + 205pp. 5⅝ x 8½.
T181 Paperbound **$1.15**

HYPOCRITICAL HELENA, Wilhelm Busch. A companion volume to "Max and Moritz," with the title piece (Die Fromme Helena) and 10 other highly amusing cartoon and verse stories, all newly translated by H. A. Klein and M. C. Klein: Adventure on New Year's Eve (Abenteuer in der Neujahrsnacht), Hangover on the Morning after New Year's Eve (Der Katzenjammer am Neujahrsmorgen), etc. English and German in parallel columns. Hours of pleasure, also a fine language aid. x + 205pp. 5⅝ x 8½. T184 Paperbound **$1.00**

THE BEAR THAT WASN'T, Frank Tashlin. What does it mean? Is it simply delightful wry humor, or a charming story of a bear who wakes up in the midst of a factory, or a satire on Big Business, or an existential cartoon-story of the human condition, or a symbolization of the struggle between conformity and the individual? New York Herald Tribune said of the first edition: ". . . a fable for grownups that will be fun for children. Sit down with the book and get your own bearings." Long an underground favorite with readers of all ages and opinions. v + 51pp. Illustrated. 5⅜ x 8½. T939 Paperbound **75¢**

RUTHLESS RHYMES FOR HEARTLESS HOMES and MORE RUTHLESS RHYMES FOR HEARTLESS HOMES, Harry Graham ("Col. D. Streamer"). Two volumes of Little Willy and 48 other poetic disasters. A bright, new reprint of oft-quoted, never forgotten, devastating humor by a precursor of today's "sick" joke school. For connoisseurs of wicked, wacky humor and all who delight in the comedy of manners. Original drawings are a perfect complement. 61 illustrations. Index. vi + 69pp. Two vols. bound as one. 5⅜ x 8½.
T930 Paperbound **75¢**

Say It language phrase books

These handy phrase books (128 to 196 pages each) make grammatical drills unnecessary for an elementary knowledge of a spoken foreign language. Covering most matters of travel and everyday life each volume contains:

Over 1000 phrases and sentences in immediately useful forms — foreign language plus English.

Modern usage designed for Americans. Specific phrases like, "Give me small change," and "Please call a taxi."

Simplified phonetic transcription you will be able to read at sight.

The only completely indexed phrase books on the market.

Covers scores of important situations: — Greetings, restaurants, sightseeing, useful expressions, etc.

These books are prepared by native linguists who are professors at Columbia, N.Y.U., Fordham and other great universities. Use them independently or with any other book or record course. They provide a supplementary living element that most other courses lack. Individual volumes in:

Russian 75¢
Hebrew 75¢
Dutch 75¢
Norwegian 75¢
Turkish 75¢
Italian 75¢
Danish 75¢
Esperanto 75¢
Polish 75¢
Spanish 75¢
Japanese 75¢
Modern Greek 75¢
French 75¢
German 75¢
Swedish 75¢
Portuguese 75¢
Yiddish 75¢
English for German-speaking people 75¢
English for Italian-speaking people 75¢
English for Spanish-speaking people 75¢

Large clear type. 128-196 pages each. 3½ x 5¼. Sturdy paper binding.

Listen and Learn language records

LISTEN & LEARN is the only language record course designed especially to meet your travel and everyday needs. It is available in separate sets for FRENCH, SPANISH, GERMAN, JAPANESE, RUSSIAN, MODERN GREEK, PORTUGUESE, ITALIAN and HEBREW, and each set contains three 33⅓ rpm long-playing records—1½ hours of recorded speech by eminent native speakers who are professors at Columbia, New York University, Queens College.

Check the following special features found only in LISTEN & LEARN:

- **Dual-language recording. 812 selected phrases and sentences,** over 3200 words, spoken first in English, then in their foreign language equivalents. A suitable pause follows each foreign phrase, allowing you time to repeat the expression. You learn by unconscious assimilation.
- **128 to 206-page manual** contains everything on the records, plus a simple phonetic pronunciation guide.
- **Indexed for convenience. The only set on the market** that is completely indexed. No more puzzling over where to find the phrase you need. Just look in the rear of the manual.
- **Practical.** No time wasted on material you can find in any grammar. LISTEN & LEARN covers central core material with phrase approach. Ideal for the person with limited learning time.
- **Living, modern expressions,** not found in other courses. Hygienic products, modern equipment, shopping—expressions used every day, like "nylon" and "air-conditioned."
- **Limited objective.** Everything you learn, no matter where you stop, is immediately useful. You have to finish other courses, wade through grammar and vocabulary drill, before they help you.
- **High-fidelity recording.** LISTEN & LEARN records equal in clarity and surface-silence any record on the market costing up to $6.

"Excellent . . . the spoken records . . . impress me as being among the very best on the market," **Prof. Mario Pei,** Dept. of Romance Languages, Columbia University. "Inexpensive and well-done . . . it would make an ideal present," CHICAGO SUNDAY TRIBUNE. "More genuinely helpful than anything of its kind which I have previously encountered," **Sidney Clark,** well-known author of "ALL THE BEST" travel books.

UNCONDITIONAL GUARANTEE. Try LISTEN & LEARN, then return it within 10 days for full refund if you are not satisfied.

Each set contains three twelve-inch 33⅓ records, manual, and album.

SPANISH	the set **$5.95**	GERMAN	the set **$5.95**
FRENCH	the set **$5.95**	ITALIAN	the set **$5.95**
RUSSIAN	the set **$5.95**	JAPANESE	the set **$6.95**
PORTUGUESE	the set **$5.95**	MODERN GREEK	the set **$5.95**
MODERN HEBREW	the set **$5.95**		

Americana

THE EYES OF DISCOVERY, J. Bakeless. A vivid reconstruction of how unspoiled America appeared to the first white men. Authentic and enlightening accounts of Hudson's landing in New York, Coronado's trek through the Southwest; scores of explorers, settlers, trappers, soldiers. America's pristine flora, fauna, and Indians in every region and state in fresh and unusual new aspects. "A fascinating view of what the land was like before the first highway went through," Time. 68 contemporary illustrations, 39 newly added in this edition. Index. Bibliography. x + 500pp. 5⅜ x 8. T761 Paperbound **$2.25**

AUDUBON AND HIS JOURNALS, J. J. Audubon. A collection of fascinating accounts of Europe and America in the early 1800's through Audubon's own eyes. Includes the Missouri River Journals —an eventful trip through America's untouched heartland, the Labrador Journals, the European Journals, the famous "Episodes", and other rare Audubon material, including the descriptive chapters from the original letterpress edition of the "Ornithological Studies", omitted in all later editions. Indispensable for ornithologists, naturalists, and all lovers of Americana and adventure. 70-page biography by Audubon's granddaughter. 38 illustrations. Index. Total of 1106pp. 5⅜ x 8.
T675 Vol I Paperbound **$2.25**
T676 Vol II Paperbound **$2.25**
The set **$4.50**

TRAVELS OF WILLIAM BARTRAM, edited by Mark Van Doren. The first inexpensive illustrated edition of one of the 18th century's most delightful books is an excellent source of first-hand material on American geography, anthropology, and natural history. Many descriptions of early Indian tribes are our only source of information on them prior to the infiltration of the white man. "The mind of a scientist with the soul of a poet," John Livingston Lowes. 13 original illustrations and maps. Edited with an introduction by Mark Van Doren. 448pp. 5⅜ x 8. T13 Paperbound **$2.00**

GARRETS AND PRETENDERS: A HISTORY OF BOHEMIANISM IN AMERICA, A. Parry. The colorful and fantastic history of American Bohemianism from Poe to Kerouac. This is the only complete record of hoboes, cranks, starving poets, and suicides. Here are Pfaff, Whitman, Crane, Bierce, Pound, and many others. New chapters by the author and by H. T. Moore bring this thorough and well-documented history down to the Beatniks. "An excellent account," N. Y. Times. Scores of cartoons, drawings, and caricatures. Bibliography. Index. xxviii + 421pp. 5⅝ x 8⅜. T708 Paperbound **$1.95**

THE EXPLORATION OF THE COLORADO RIVER AND ITS CANYONS, J. W. Powell. The thrilling first-hand account of the expedition that filled in the last white space on the map of the United States. Rapids, famine, hostile Indians, and mutiny are among the perils encountered as the unknown Colorado Valley reveals its secrets. This is the only uncut version of Major Powell's classic of exploration that has been printed in the last 60 years. Includes later reflections and subsequent expedition. 250 illustrations, new map. 400pp. 5⅝ x 8⅜. T94 Paperbound **$2.25**

THE JOURNAL OF HENRY D. THOREAU, Edited by Bradford Torrey and Francis H. Allen. Henry Thoreau is not only one of the most important figures in American literature and social thought; his voluminous journals (from which his books emerged as selections and crystallizations) constitute both the longest, most sensitive record of personal internal development and a most penetrating description of a historical moment in American culture. This present set, which was first issued in fourteen volumes, contains Thoreau's entire journals from 1837 to 1862, with the exception of the lost years which were found only recently. We are reissuing it, complete and unabridged, with a new introduction by Walter Harding, Secretary of the Thoreau Society. Fourteen volumes reissued in two volumes. Foreword by Henry Seidel Canby. Total of 1888pp. 8⅜ x 12¼. T312-3 Two volume set, Clothbound **$20.00**

GAMES AND SONGS OF AMERICAN CHILDREN, collected by William Wells Newell. A remarkable collection of 190 games with songs that accompany many of them; cross references to show similarities, differences among them; variations; musical notation for 38 songs. Textual discussions show relations with folk-drama and other aspects of folk tradition. Grouped into categories for ready comparative study: Love-games, histories, playing at work, human life, bird and beast, mythology, guessing-games, etc. New introduction covers relations of songs and dances to timeless heritage of folklore, biographical sketch of Newell, other pertinent data. A good source of inspiration for those in charge of groups of children and a valuable reference for anthropologists, sociologists, psychiatrists. Introduction by Carl Withers. New indexes of first lines, games. 5⅜ x 8½. xii + 242pp. T354 Paperbound **$1.75**

Art, History of Art, Antiques, Graphic Arts, Handcrafts

ART STUDENTS' ANATOMY, E. J. Farris. Outstanding art anatomy that uses chiefly living objects for its illustrations. 71 photos of undraped men, women, children are accompanied by carefully labeled matching sketches to illustrate the skeletal system, articulations and movements, bony landmarks, the muscular system, skin, fasciae, fat, etc. 9 x-ray photos show movement of joints. Undraped models are shown in such actions as serving in tennis, drawing a bow in archery, playing football, dancing, preparing to spring and to dive. Also discussed and illustrated are proportions, age and sex differences, the anatomy of the smile, etc. 8 plates by the great early 18th century anatomic illustrator Siegfried Albinus are also included. Glossary. 158 figures, 7 in color. x + 159pp. 5⅝ x 8⅜. T744 Paperbound **$1.50**

AN ATLAS OF ANATOMY FOR ARTISTS, F Schider. A new 3rd edition of this standard text enlarged by 52 new illustrations of hands, anatomical studies by Cloquet, and expressive life studies of the body by Barcsay. 189 clear, detailed plates offer you precise information of impeccable accuracy. 29 plates show all aspects of the skeleton, with closeups of special areas, while 54 full-page plates, mostly in two colors, give human musculature as seen from four different points of view, with cutaways for important portions of the body. 14 full-page plates provide photographs of hand forms, eyelids, female breasts, and indicate the location of muscles upon models. 59 additional plates show how great artists of the past utilized human anatomy. They reproduce sketches and finished work by such artists as Michelangelo, Leonardo da Vinci, Goya, and 15 others. This is a lifetime reference work which will be one of the most important books in any artist's library. "The standard reference tool," AMERICAN LIBRARY ASSOCIATION. "Excellent," AMERICAN ARTIST. Third enlarged edition. 189 plates, 647 illustrations. xxvi + 192pp. 7⅞ x 10⅝. T241 Clothbound **$6.00**

AN ATLAS OF ANIMAL ANATOMY FOR ARTISTS, W. Ellenberger, H. Baum, H. Dittrich. The largest, richest animal anatomy for artists available in English. 99 detailed anatomical plates of such animals as the horse, dog, cat, lion, deer, seal, kangaroo, flying squirrel, cow, bull, goat, monkey, hare, and bat. Surface features are clearly indicated, while progressive beneath-the-skin pictures show musculature, tendons, and bone structure. Rest and action are exhibited in terms of musculature and skeletal structure and detailed cross-sections are given for heads and important features. The animals chosen are representative of specific families so that a study of these anatomies will provide knowledge of hundreds of related species. "Highly recommended as one of the very few books on the subject worthy of being used as an authoritative guide," DESIGN. "Gives a fundamental knowledge," AMERICAN ARTIST. Second revised, enlarged edition with new plates from Cuvier, Stubbs, etc. 288 illustrations. 153pp. 11⅜ x 9. T82 Clothbound **$6.00**

THE HUMAN FIGURE IN MOTION, Eadweard Muybridge. The largest selection in print of Muybridge's famous high-speed action photos of the human figure in motion. 4789 photographs illustrate 162 different actions: men, women, children—mostly undraped—are shown walking, running, carrying various objects, sitting, lying down, climbing, throwing, arising, and performing over 150 other actions. Some actions are shown in as many as 150 photographs each. All in all there are more than 500 action strips in this enormous volume, series shots taken at shutter speeds of as high as 1/6000th of a second! These are not posed shots, but true stopped motion. They show bone and muscle in situations that the human eye is not fast enough to capture. Earlier, smaller editions of these prints have brought $40 and more on the out-of-print market. "A must for artists," ART IN FOCUS. "An unparalleled dictionary of action for all artists," AMERICAN ARTIST. 390 full-page plates, with 4789 photographs. Printed on heavy glossy stock. Reinforced binding with headbands. xxi + 390pp. 7⅞ x 10⅝. T204 Clothbound **$10.00**

ANIMALS IN MOTION, Eadweard Muybridge. This is the largest collection of animal action photos in print. 34 different animals (horses, mules, oxen, goats, camels, pigs, cats, guanacos, lions, gnus, deer, monkeys, eagles—and 21 others) in 132 characteristic actions. The horse alone is shown in more than 40 different actions. All 3919 photographs are taken in series at speeds up to 1/6000th of a second. The secrets of leg motion, spinal patterns, head movements, strains and contortions shown nowhere else are captured. You will see exactly how a lion sets his foot down; how an elephant's knees are like a human's—and how they differ; the position of a kangaroo's legs in mid-leap; how an ostrich's head bobs; details of the flight of birds—and thousands of facets of motion only the fastest cameras can catch. Photographed from domestic animals and animals in the Philadelphia zoo, it contains neither semiposed artificial shots nor distorted telephoto shots taken under adverse conditions. Artists, biologists, decorators, cartoonists, will find this book indispensable for understanding animals in motion. "A really marvelous series of plates," NATURE (London). "The dry plate's most spectacular early use was by Eadweard Muybridge," LIFE. 3919 photographs; 380 full pages of plates. 440pp. Printed on heavy glossy paper. Deluxe binding with headbands. 7⅞ x 10⅝. T203 Clothbound **$10.00**

THE AUTOBIOGRAPHY OF AN IDEA, Louis Sullivan. The pioneer architect whom Frank Lloyd Wright called "the master" reveals an acute sensitivity to social forces and values in this passionately honest account. He records the crystallization of his opinions and theories, the growth of his organic theory of architecture that still influences American designers and architects, contemporary ideas, etc. This volume contains the first appearance of 34 full-page plates of his finest architecture. Unabridged reissue of 1924 edition. New introduction by R. M. Line. Index. xiv + 335pp. 5⅜ x 8. T281 Paperbound **$2.00**

THE DRAWINGS OF HEINRICH KLEY. The first uncut republication of both of Kley's devastating sketchbooks, which first appeared in pre-World War I Germany. One of the greatest cartoonists and social satirists of modern times, his exuberant and iconoclastic fantasy and his extraordinary technique place him in the great tradition of Bosch, Breughel, and Goya, while his subject matter has all the immediacy and tension of our century. 200 drawings. viii + 128pp. 7¾ x 10¾. T24 Paperbound **$1.85**

MORE DRAWINGS BY HEINRICH KLEY. All the sketches from Leut' Und Viecher (1912) and Sammel-Album (1923) not included in the previous Dover edition of Drawings. More of the bizarre, mercilessly iconoclastic sketches that shocked and amused on their original publication. Nothing was too sacred, no one too eminent for satirization by this imaginative, individual and accomplished master cartoonist. A total of 158 illustrations. Iv + 104pp. 7¾ x 10¾. T41 Paperbound **$1.85**

PINE FURNITURE OF EARLY NEW ENGLAND, R. H. Kettell. A rich understanding of one of America's most original folk arts that collectors of antiques, interior decorators, craftsmen, woodworkers, and everyone interested in American history and art will find fascinating and immensely useful. 413 illustrations of more than 300 chairs, benches, racks, beds, cupboards, mirrors, shelves, tables, and other furniture will show all the simple beauty and character of early New England furniture. 55 detailed drawings carefully analyze outstanding pieces. "With its rich store of illustrations, this book emphasizes the individuality and varied design of early American pine furniture. It should be welcomed," ANTIQUES. 413 illustrations and 55 working drawings. 475. 8 x 10¾. T145 Clothbound **$10.00**

THE HUMAN FIGURE, J. H. Vanderpoel. Every important artistic element of the human figure is pointed out in minutely detailed word descriptions in this classic text and illustrated as well in 430 pencil and charcoal drawings. Thus the text of this book directs your attention to all the characteristic features and subtle differences of the male and female (adults, children, and aged persons), as though a master artist were telling you what to look for at each stage. 2nd edition, revised and enlarged by George Bridgman. Foreword. 430 illustrations. 143pp. 6⅛ x 9¼. T432 Paperbound **$1.50**

LETTERING AND ALPHABETS, J. A. Cavanagh. This unabridged reissue of LETTERING offers a full discussion, analysis, illustration of 89 basic hand lettering styles — styles derived from Caslons, Bodonis, Garamonds, Gothic, Black Letter, Oriental, and many others. Upper and lower cases, numerals and common signs pictured. Hundreds of technical hints on make-up, construction, artistic validity, strokes, pens, brushes, white areas, etc. May be reproduced without permission! 89 complete alphabets; 72 lettered specimens. 121pp. 9⅜ x 8. T53 Paperbound **$1.35**

STICKS AND STONES, Lewis Mumford. A survey of the forces that have conditioned American architecture and altered its forms. The author discusses the medieval tradition in early New England villages; the Renaissance influence which developed with the rise of the merchant class; the classical influence of Jefferson's time; the "Mechanicsvilles" of Poe's generation; the Brown Decades; the philosophy of the Imperial facade; and finally the modern machine age. "A truly remarkable book," SAT. REV. OF LITERATURE. 2nd revised edition. 21 illustrations. xvii + 228pp. 5⅜ x 8. T202 Paperbound **$1.75**

THE STANDARD BOOK OF QUILT MAKING AND COLLECTING, Marguerite Ickis. A complete easy-to-follow guide with all the information you need to make beautiful, useful quilts. How to plan, design, cut, sew, appliqué, avoid sewing problems, use rag bag, make borders, tuft, every other aspect. Over 100 traditional quilts shown, including over 40 full-size patterns. At-home hobby for fun, profit. Index. 483 illus. 1 color plate. 287pp. 6¾ x 9½. T582 Paperbound **$2.00**

THE BOOK OF SIGNS, Rudolf Koch. Formerly $20 to $25 on the out-of-print market, now only $1.00 in this unabridged new edition! 493 symbols from ancient manuscripts, medieval cathedrals, coins, catacombs, pottery, etc. Crosses, monograms of Roman emperors, astrological, chemical, botanical, runes, housemarks, and 7 other categories. Invaluable for handicraft workers, illustrators, scholars, etc., this material may be reproduced without permission. 493 illustrations by Fritz Kredel. 104pp. 6½ x 9¼. T162 Paperbound **$1.00**

PRIMITIVE ART, Franz Boas. This authoritative and exhaustive work by a great American anthropologist covers the entire gamut of primitive art. Pottery, leatherwork, metal work, stone work, wood, basketry, are treated in detail. Theories of primitive art, historical depth in art history, technical virtuosity, unconscious levels of patterning, symbolism, styles, literature, music, dance, etc. A must book for the interested layman, the anthropologist, artist, handicrafter (hundreds of unusual motifs), and the historian. Over 900 illustrations (50 ceramic vessels, 12 totem poles, etc.). 376pp. 5⅜ x 8. T25 Paperbound **$2.25**

Fiction

FLATLAND, E. A. Abbott. A science-fiction classic of life in a 2-dimensional world that is also a first-rate introduction to such aspects of modern science as relativity and hyperspace. Political, moral, satirical, and humorous overtones have made FLATLAND fascinating reading for thousands. 7th edition. New introduction by Banesh Hoffmann. 16 illustrations. 128pp. 5⅜ x 8. T1 Paperbound **$1.00**

THE WONDERFUL WIZARD OF OZ, L. F. Baum. Only edition in print with all the original W. W. Denslow illustrations in full color—as much a part of "The Wizard" as Tenniel's drawings are of "Alice in Wonderland." "The Wizard" is still America's best-loved fairy tale, in which, as the author expresses it, "The wonderment and joy are retained and the heartaches and nightmares left out." Now today's young readers can enjoy every word and wonderful picture of the original book. New introduction by Martin Gardner. A Baum bibliography. 23 full-page color plates. viii + 268pp. 5⅜ x 8. T691 Paperbound **$1.50**

THE MARVELOUS LAND OF OZ, L. F. Baum. This is the equally enchanting sequel to the "Wizard," continuing the adventures of the Scarecrow and the Tin Woodman. The hero this time is a little boy named Tip, and all the delightful Oz magic is still present. This is the Oz book with the Animated Saw-Horse, the Woggle-Bug, and Jack Pumpkinhead. All the original John R. Neill illustrations, 10 in full color. 287 pp. 5⅜ x 8. T692 Paperbound **$1.50**

28 SCIENCE FICTION STORIES OF H. G. WELLS. Two full unabridged novels, MEN LIKE GODS and STAR BEGOTTEN, plus 26 short stories by the master science-fiction writer of all time! Stories of space, time, invention, exploration, future adventure—an indispensable part of the library of everyone interested in science and adventure. PARTIAL CONTENTS: Men Like Gods, The Country of the Blind, In the Abyss, The Crystal Egg, The Man Who Could Work Miracles, A Story of the Days to Come, The Valley of Spiders, and 21 more! 928pp. 5⅜ x 8.
T265 Clothbound **$4.50**

THREE MARTIAN NOVELS, Edgar Rice Burroughs. Contains: Thuvia, Maid of Mars; The Chessmen of Mars; and The Master Mind of Mars. High adventure set in an imaginative and intricate conception of the Red Planet. Mars is peopled with an intelligent, heroic human race which lives in densely populated cities and with fierce barbarians who inhabit dead sea bottoms. Other exciting creatures abound amidst an inventive framework of Martian history and geography. Complete unabridged reprintings of the first edition. 16 illustrations by J. Allen St. John. vi + 499pp. 5⅜ x 8½. T39 Paperbound **$1.85**

SEVEN SCIENCE FICTION NOVELS, H. G. Wells. Full unabridged texts of 7 science-fiction novels of the master. Ranging from biology, physics, chemistry, astronomy to sociology and other studies, Mr. Wells extrapolates whole worlds of strange and intriguing character. "One will have to go far to match this for entertainment, excitement, and sheer pleasure . . . ," NEW YORK TIMES. Contents: The Time Machine, The Island of Dr. Moreau, First Men in the Moon, The Invisible Man, The War of the Worlds, The Food of the Gods, In the Days of the Comet. 1015pp. 5⅜ x 8. T264 Clothbound **$4.50**

THE LAND THAT TIME FORGOT and THE MOON MAID, Edgar Rice Burroughs. In the opinion of many, Burroughs' best work. The first concerns a strange island where evolution is individual rather than phylogenetic. Speechless anthropoids develop into intelligent human beings within a single generation. The second projects the reader far into the future and describes the first voyage to the Moon (in the year 2025), the conquest of the Earth by the Moon, and years of violence and adventure as the enslaved Earthmen try to regain possession of their planet. "An imaginative tour de force that keeps the reader keyed up and expectant," NEW YORK TIMES. Complete, unabridged text of the original two novels (three parts in each). 5 illustrations by J. Allen St. John. vi + 552pp. 5⅜ x 8½.
T1020 Clothbound **$3.75**
T358 Paperbound **$2.00**

3 ADVENTURE NOVELS by H. Rider Haggard. Complete texts of "She," "King Solomon's Mines," "Allan Quatermain." Qualities of discovery; desire for immortality; search for primitive, for what is unadorned by civilization, have kept these novels of African adventure exciting, alive to readers from R. L. Stevenson to George Orwell. 636pp. 5⅜ x 8.
T584 Paperbound **$2.00**

A PRINCESS OF MARS and A FIGHTING MAN OF MARS: TWO MARTIAN NOVELS BY EDGAR RICE BURROUGHS. "Princess of Mars" is the very first of the great Martian novels written by Burroughs, and it is probably the best of them all; it set the pattern for all of his later fantasy novels and contains a thrilling cast of strange peoples and creatures and the formula of Olympian heroism amidst ever-fluctuating fortunes which Burroughs carries off so successfully. "Fighting Man" returns to the same scenes and cities—many years later. A mad scientist, a degenerate dictator, and an indomitable defender of the right clash—with the fate of the Red Planet at stake! Complete, unabridged reprinting of original editions. Illustrations by F. E. Schoonover and Hugh Hutton. v + 356pp. 5⅜ x 8½.
T1140 Paperbound **$1.75**

Music

A GENERAL HISTORY OF MUSIC, Charles Burney. A detailed coverage of music from the Greeks up to 1789, with full information on all types of music: sacred and secular, vocal and instrumental, operatic and symphonic. Theory, notation, forms, instruments, innovators, composers, performers, typical and important works, and much more in an easy, entertaining style. Burney covered much of Europe and spoke with hundreds of authorities and composers so that this work is more than a compilation of records . . . it is a living work of careful and first-hand scholarship. Its account of thoroughbass (18th century) Italian music is probably still the best introduction on the subject. A recent NEW YORK TIMES review said, "Surprisingly few of Burney's statements have been invalidated by modern research . . . still of great value." Edited and corrected by Frank Mercer. 35 figures. Indices. 1915pp. 5⅜ x 8. 2 volumes. T36 The Set, Clothbound **$12.50**

A DICTIONARY OF HYMNOLOGY, John Julian. This exhaustive and scholarly work has become known as an invaluable source of hundreds of thousands of important and often difficult to obtain facts on the history and use of hymns in the western world. Everyone interested in hymns will be fascinated by the accounts of famous hymns and hymn writers and amazed by the amount of practical information he will find. More than 30,000 entries on individual hymns, giving authorship, date and circumstances of composition, publication, textual variations, translations, denominational and ritual usage, etc. Biographies of more than 9,000 hymn writers, and essays on important topics such as Christmas carols and children's hymns, and much other unusual and valuable information. A 200 page double-columned index of first lines — the largest in print. Total of 1786 pages in two reinforced clothbound volumes. 6¼ x 9¼. The set, T333 Clothbound **$17.50**

MUSIC IN MEDIEVAL BRITAIN, F. Ll. Harrison. The most thorough, up-to-date, and accurate treatment of the subject ever published, beautifully illustrated. Complete account of institutions and choirs; carols, masses, and motets; liturgy and plainsong; and polyphonic music from the Norman Conquest to the Reformation. Discusses the various schools of music and their reciprocal influences; the origin and development of new ritual forms; development and use of instruments; and new evidence on many problems of the period. Reproductions of scores, over 200 excerpts from medieval melodies. Rules of harmony and dissonance; influence of Continental styles; great composers (Dunstable, Cornysh, Fairfax, etc.); and much more. Register and index of more than 400 musicians. Index of titles. General Index. 225-item bibliography. 6 Appendices. xix + 491pp. 5⅝ x 8¾. T705 Clothbound **$10.00**

THE MUSIC OF SPAIN, Gilbert Chase. Only book in English to give concise, comprehensive account of Iberian music; new Chapter covers music since 1941. Victoria, Albéniz, Cabezón, Pedrell, Turina, hundreds of other composers; popular and folk music; the Gypsies; the guitar; dance, theatre, opera, with only extensive discussion in English of the Zarzuela; virtuosi such as Casals; much more. "Distinguished . . . readable," Saturday Review. 400-item bibliography. Index. 27 photos. 383pp. 5⅜ x 8. T549 Paperbound **$2.25**

ON STUDYING SINGING, Sergius Kagen. An intelligent method of voice-training, which leads you around pitfalls that waste your time, money, and effort. Exposes rigid, mechanical systems, baseless theories, deleterious exercises. "Logical, clear, convincing . . . dead right," Virgil Thomson, N.Y. Herald Tribune. "I recommend this volume highly," Maggie Teyte, Saturday Review. 119pp. 5⅜ x 8. T622 Paperbound **$1.35**

Prices subject to change without notice.

Dover publishes books on art, music, philosophy, literature, languages, history, social sciences, psychology, handcrafts, orientalia, puzzles and entertainments, chess, pets and gardens, books explaining science, intermediate and higher mathematics, mathematical physics, engineering, biological sciences, earth sciences, classics of science, etc. Write to:

Dept. catrr.
Dover Publications, Inc.
180 Varick Street, N.Y. 14, N.Y.